AF731051

HERBERT D. SMITH
and the
CITRUS BLACKFLY

Herbert D. Smith
and the
Citrus Blackfly

David Lewis

Based on the Reports and Letters of
Herbert D. Smith

‘

Shaw Press

Shaw Press
PO Box 20070
San Jose, California 95160
www.shawpress.com

Copyright © 2016 by David C. Lewis
Herbert D. Smith letters © 2016 Estate of Herbert D. Smith

Dedicated to Herbert D. Smith
and all scientists working to improve our lives,
and to David S. Lewis, whose own life
of commitment and service could fill a book

CONTENTS

Preface

Herbert Smith was an old-school scientist. He believed in science as a method to solve real world problems. He wasn't much interested in titles, awards, conferences at fancy hotels on the beach in Hawaii, publishing paper after paper, or working on never-ending projects that provided steady employment but never yielded any useful result. He believed in hard work – digging in and getting his hands dirty. He believed in getting the job done. Herb was an entomologist. The job he undertook was to try to save the North American citrus industry from destruction by an insect pest known as the citrus blackfly.

I knew of Herbert Smith from an early age, though I never met him. He was my great uncle. It was through my mother's childhood stamp collection that I first got some dim awareness of this relative who spent a lifetime traveling the world. There were colorful stamps in many languages from exotic countries: the speeding train commemorating the 1937 Paris Congrés des Chemin de Fer, the Malayan stamp with King George among the palm trees, the old high-winged Ford Trimotor airplane flying into the Republica de Cuba, the many armed dancing Nataraja on the stamp from newly independent India.

Visiting my grandparents a few years later, I discovered a binder, compiled by Herb's daughter Lucienne, containing copies of dozens of letters that the well-traveled scientist had written home to his family. The letters were a remarkable collection that revealed a life of both great adventure and great sacrifice. Every few years I would pull out the binder and reread them. I was struck by the varying paths a life could take. Herb, by then a middle-aged man, was writing home about his life in remote areas of the world, traveling by ox cart, sleeping on a bedroll on the floor of train stations and rundown inns, picking flies out of his food, missing the graduations and weddings of his children, missing the births of his grandchildren. That was a far different life than my gradparents had, or anyone else I knew.

A few years ago I traveled with my own family to meet Herb's daughter Janet, and to learn more about her father. Our little rented Renault wound its way up Impressionist Hill on the outskirts of Paris to the village of Bougival, then through the gate and up the gravel drive of the home of Janet and her husband Jean-Jacques. Janet is more French than American, having lived in France for most of her childhood and all of her adult life after college. Jean-Jacques showed us his well-tended garden with his prized red and white currant bushes, and later at dinner regaled us with fascinating stories of his own storied life, growing up in the very different Paris of the 1920s and 30s, surviving the brutal war years and occupation, and building a business and raising a family in the years after the war's end.

But what I most wanted to learn about was Herbert Smith. Janet generously shared details about his early life in the States, life on the French Riviera and Paris in the years prior to the outbreak of war in Europe, and Herb's later years working in the field in Latin America and Asia.

"My father had a hard life," Janet told me. Indeed, the hard life of Herbert Smith led to many sacrifices for both himself and his family. But one standard for measuring a life is how that life affected others. The story that unfolded through Herb Smith's research papers and personal letters shows that his work benefited countless others. That is a successful life in my book.

Entomologist Herbert D. Smith working in the field in Mexico to control the citrus blackfly with parasites he collected across Asia (Smith family collection).

1

Mexico: A New Danger

In June 1943 an old Ford truck turned off the Los Mochis-Culiacan Road at Pericos and headed northeast. The truck creaked and jolted up the dusty dirt road and after a day's journey from Culiacan pulled into the central square of the little town of Badiraguato. It was fortunate that the heavy mid-summer rains had not yet arrived, as they make the Badiraguato road impassably muddy, requiring several days on horseback to reach the town.

A bespectacled middle-aged American got out and looked around. Badiraguato saw few foreigners passing through, for it was an impoverished Mexican backwater well off the tourist routes. The American was U.S. Department of Agriculture entomologist Herbert Smith. He was accompanied by a Mexican agronomist, Alfonso Delgado de Garay. They were welcomed by the mayor Miguel Sandoval, who had requested their assistance. Something was making leaves and fruit turn black in the citrus groves

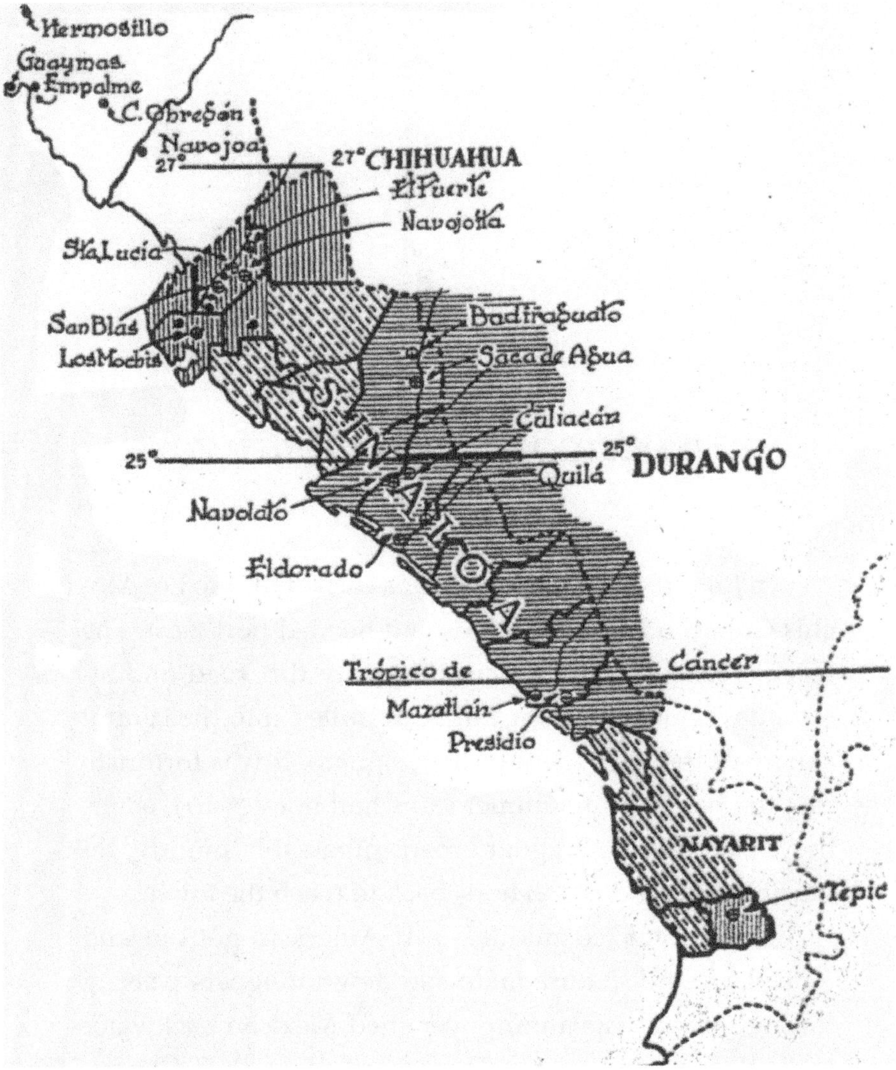

In 1935 the citrus blackfly, a devastating pest of citrus trees, was discovered at Eldorado, Mexico. It gradually spread north and south across Sinaloa and into Nayarit. In 1943 Herbert Smith brought *Eretmocerous serius* parasites to Mexico and with the cooperation of the Mexican agricultural departments released them into infested areas, including the locations on this map.

scattered around the hills east of the town, not to mention in the mayor's own garden.

The problem wasn't limited to Badiraguato. In the years of World War II a new battle line was forming south of the American border. Mexico had declared war on Germany in 1942 after two of its tankers were torpedoed off the Florida coast. The conflict though was in name only, without real fighting. There was however a new enemy on the home front that was doing real damage, a formidable insect pest of citrus trees.

In 1935, citrus growers at Eldorado in the Mexican state of Sinaloa began to notice a dark, sooty coating on many leaves. The infestation was identified as the citrus blackfly (*Aleurocanthus woglumi*). The citrus blackfly causes damage by reproducing prolifically on the bottom of citrus fruit tree leaves. The trees become weak and may even die. The flies produce large amounts of a sticky material called honeydew which in turn allows a sooty-mold fungus to grow that blackens the leaves and fruit.

This pest had not been hitherto seen in Mexico, but its appearance was not exactly a surprise. The citrus blackfly is native to Asia. In 1910 it was first described by three different entomologists traveling around Asia. Harold Maxwell-Lefroy found it in India and George Compere found it in Manila. Russell Woglum reported its occurrence in a variety of locations around India.

By this time, however, it appears the citrus blackfly had already made its way to the New World, as in 1914 S. F. Ashby collected specimens in Jamaica and reported it was already widespread there. In 1916 Patricio Cardin found the blackfly in Cuba as did Lewis J. K. Brace at Nassau in the Bahamas. The next year Harold Morrison noted it was also established in the Canal Zone and in parts of Panama. The blackfly was spreading rapidly around the Caribbean.

The great increase in world travel in the modern era has had the unfortunate side effect of allowing insect pests to hitch rides to new lands. Even more unfortunate, other insect species that had evolved over thousands of years to live off the pest in a parasitic or predatory fashion, thereby keeping pest populations at a tolerable level, often failed to make the trip. The result tended to be an explosion of pests which had free reign to go to town in their new environment.

Scientists distinguish two types of animals that feed on other animals. Predators typically consume multiple species of animals smaller than themselves. Parasites are usually smaller than their host, often evolved to live on one species or a few closely related species. Most parasites allow the host to survive. In the insect world some parasites live on very small insects and as a result the parasites can be so tiny as to be difficult to even see with naked eye.

A small group of scientists within the U. S. Department of Agriculture sought to use parasites to try to reign in the citrus blackfly. Biological control was promising as the citrus blackfly was a minor pest in Asia, unlike its aggressive and destructive appearance in the Caribbean. There must be other insects that were limiting the population of the fly in Asia. If these predator or parasites could be found and reunited with the blackfly, control might

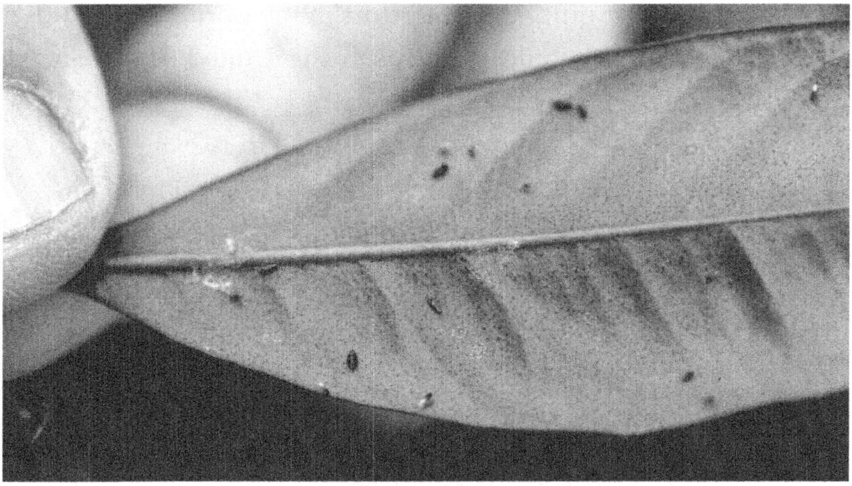

The citrus blackfly in various stages of life on citrus leaves. (Photos courtesy of Dr. Ru Nguyen, Florida Department of Agriculture, Gainesville, Florida.)

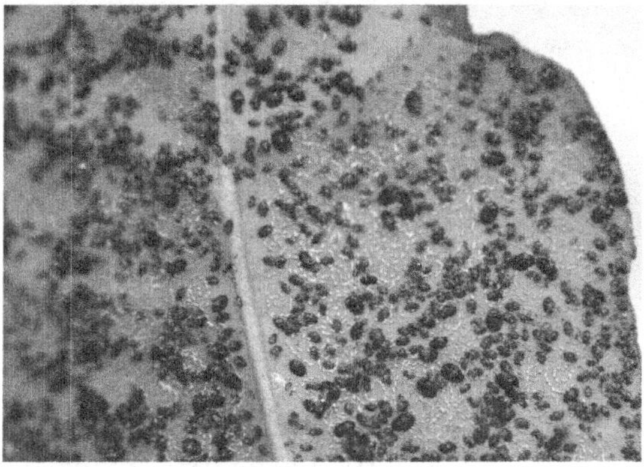

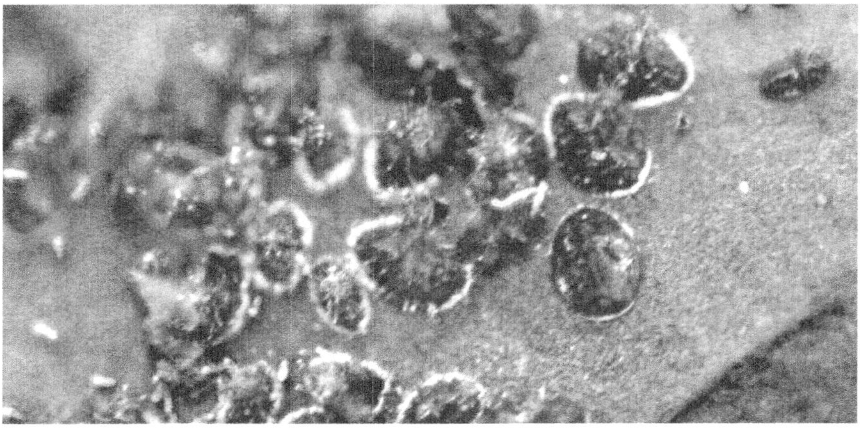

be achieved. Failure of their effort in the Caribbean could mean the pest would soon reach the United States and devastate the citrus industry there.

Over a three year period beginning in 1929 a parasitic wasp called *Eretmocerus serius* Silv. was located in Malaya by Department of Agriculture entomologist Curtis Clausen and brought over and introduced into infested citrus groves in Cuba. This wasp reproduces by laying its eggs inside the blackfly larva. This is good for the wasp but fatal for the blackfly. Within nine months of release, infested citrus groves were free from significant infestation and had returned to good health. The process was repeated with similar good results in Jamaica, Haiti, the Bahamas, the Canal Zone, and Costa Rica.

The citrus blackfly infestations in Mexico spread north and south from Sinaloa, becoming serious by 1940. Later that year the pest jumped hundreds of miles to the south, where it was found in a lime grove in Colima state. By 1942 the affected areas extended though the coastal areas of Nayarit, Colima and Jalisco states.

In July 1942 the citrus blackfly problem was a major point of discussion at the Second Inter-American Conference on Agriculture in Mexico City. The Mexican Department of Agriculture requested assistance from the Bureau of Entomology and Quarantine of the counterpart American agency. The Americans had their own reasons to want to be involved, as the uncontrolled spread of the citrus blackfly across Mexico put the American citrus industry in great peril.

By the early 1940s the US citrus industry had grown tremendously with a crop worth $400 million a year, the equivalent of about seven billion dollars today. The annual production was seven million tons of fruit. American agricultural interests were keeping a watchful eye on developments in Mexico. There were 100,000 acres of citrus trees under cultivation in Texas, mostly in the Rio Grande Valley, a stone's throw from the Mexican border.

In 1943 U.S. Department of Agriculture entomologist Herbert Smith undertook the task of trying to stop the epidemic in Mexico, as had been done earlier in the Caribbean. Herb was one of the department's most experienced scientists, having worked in the biological control field for more than twenty years. Smith had spent many of those years manning an American laboratory in France, studying native parasitic species that might control pests that had made their way from Europe to the United States. The outbreak of war on the Continent led to the evacuation of the European lab, and Smith was now available for work in Latin America.

Smith left New York by train in early May bound for Mexico City. Once in the capitol, Smith met with Dario L. Arrieta M., the Director General of the Direccion de Agricultura; Ricardo Coronado Padilla, Chief of the Departmento de Fitosanitario; and Alfonso Delgado de Gary, Chief of the Seccion de Investigacion y Divulgacion.

Smith headed for the West Coast of Mexico on May 22nd, carrying a letter of introduction from Arrieta requesting all cooperation from the police and military authorities in Colima, Jalisco and Sinaloa. On a stop in Guadalajara to change trains, he met with Alvaro Peniche Canto who was in charge of pest control for Colima, Jalisco and Nayarit.

Arriving in Mazatlan, Smith tried to meet up with the State Chief of Fitosanitario, but the latter was unavailable. As a result Herb was unable to get a vehicle to use for surveys that had been promised to him. He was able to make arrangements there with the local Pan Am agent to take receipt of parasite shipments and to store them in a refrigerator until he could retrieve them.

Smith continued on to Culiacan where he called on the Mexican agricultural official for Sinaloa, Enrique Coppel Rivas. Rivas was very knowledgeable about the blackfly situation in Sinaloa, and offered Smith the use of one of his trucks to do surveys and parasite colonization there.

On May 27th Herb traveled by truck to the large sugar plantation at El Dorado where the blackfly was first noticed in Mexico in 1935. The plantation was owned by the prominent Redo family, who had arranged at the turn of the century for a complete sugar mill to be transported from Florida around the Great Horn and thence overland to El Dorado. Diego Redo welcomed Herb there and offered the assistance of his own private entomologist Temistocles Voglioti. The blackfly infestation at El Dorado was heavy and widespread.

On June 6th Alfonso Delgado arrived from Mexico City bringing the first shipment of parasites that had been received from Panama. A week later the first releases were made at Badiraguato where Mayor Sandoval had asked for help from the heavy infestations of orange trees in the mountains east of town,

In mid-June Smith surveyed the area around Colima with the assistance of several local officials. The pest was in the town, but had not spread far. Smith and Delgado made numerous releases of *Eretmocerus serius* in Colima onto citrus trees infested with the citrus blackfly. Twenty were released at the entrance to the Hacienda La Albarradita. Over a hundred parasites were each placed in the garden of Senorita Rosario Fernandez on Calle Gabino Barreda and at a notary office a kilometer from the train station. A large release of 347 was made at La Florida on the road to the station. Twenty other groups of parasites were released at various haciendas and orchards in the area, and locations and contacts were entered into Smith's detailed notebook so he could return to monitor the results.

Next Smith traveled to northern Sinaloa to assess the situation there. Los Mochis was clean of blackfly but the pest was causing moderate infestations at San Blas. At Guaymas in Sonora State no blackfly was found. Empalme, Hermosillo, Ciudad Obregon and Navojoa were also free of the pest.

Delgado meanwhile went to Tepic, Nayarit, where he made three releases totaling 317 parasites into nearby orchards. There were two more releases of 444 bugs at Presidio, Sinaloa, into lemon groves there. The list of releases goes on and on: four in Mazatlan (422 parasites), 25 bugs on a lemon tree at the restaurant next to the station at Quila, 9 releases around El Dorado, three at La Aurora near

Culiacan, one at Navolato, one at Saca de Agua, and three at El Fuerte. Then there were many more as he traveled to Santa Lucia, Navojoita, San Blas, and Los Mochis.

In all Smith and Delgado made 60 releases into lemon and limes groves. A total of 8,767 *Eretmocerus serius* parasites were set out to hopefully start parasitizing the heavy infestations of the citrus blackfly, and bringing the trees back to good health. Only time would tell.

2

U.S.: To Be an Entomologist

As far as is known Herbert Dwight Smith wasn't one of those boys with a passion for bugs and such creatures, but while working summers on nearby farms he had developed an appreciation of the battle with insects that farmers faced every day. Entomology was a field that held great promise to benefit American agriculture.

Herb was born in New Haven, Connecticut on June 29, 1894, the youngest of four children. His father Wilford Smith was a woodworker who seemed to move every few years, spending time in Glenville, Nebraska; Saginaw, Michigan; New Haven; Walpole, Massachusetts; Toledo, Ohio; Titusville, Pennsylvania; and Poughkeepsie, New York. Herbert had a brother Leon and two sisters Edith and Ruth. His mother Elizabeth, known as "Lizy," died when he was 13.

Smith was a smart boy and knew how to tell a good story. Years later Herb's family made an interesting dis-

Herbert Smith earned a degree in Entomology at the Massachusetts Agricultural College at Amherst in 1917.

An early photo of Smith studying insects at his microscope (Smith family collection).

covery inside his father's old business ledger listing his carpentry notes and accounts. It was a short autobiography that Herb had written as a young boy titled "The Life of Herbert Smith." In a section he called "Life of Hero" Herb detailed moving from town to town – in Walpole he played –- in Toledo he was "not a very good boy" –- in Titusville he was "a good scholar" –- in Poughkeepsie he was happy. He concluded with the words "then ??" to indicate his unknown future. It turned out life would take him to many adventures far away from Poughkeepsie.

After graduating from high school in Poughkeepsie Smith attended the Massachusetts Agricultural College at Amherst (now a campus of the University of Massachusetts). He received his B.S. degree in Entomology there in 1917.

Entomology as a formal academic field of study was comparatively new at that time, but humans have a long history of dealing with insects. Insects are by most measures the most successful group of animals on the planet. The million known species of insects make up more than half of all known animal life forms. Since it is estimated that an additional 5 to 10 million yet undescribed insect species may be living on the earth, insects may actually make up 90% of all the earth's animal lifeforms.

A majority of insects go about their lives and have little impact on human populations. Some species serve important ecological roles that are desirable to humans, such as the pollination of flowering plants. There are also many species of insects, however, that are considered to be pests, causing extensive damage to crops or spreading human disease.

For thousands of years farmers have noticed that some insects can control other insects on plants. The Chinese were the first to use natural enemies to control insect pests. Chinese historical records indicate that in the third century, nests of an ant now known as *Oecophylla smaragdina* were sold in southern China to control *Tesseratoma papillosa* on citrus plants. Similarly ant nests were used in 13th century Yemen to control pests of the date palm. The ability of ladybird beetles to control aphids and scales has been known for 800 years.

A number of European scientists in the 17th and 18th century noticed cases of biological control of insect species by another insect or by non-insect species such as birds and fungi. The hated bedbug, *Cimex lectularius*, was treated by releasing a pentatomid predator called *Picromerus bidens*. Importing mynah birds to Mauritius helped control the red locust, *Nomadacris septemfasciata*.

Scientific research became more systemic in the 19th century. Darwin proposed using ichneumon wasps for biological control of cabbage caterpillars. The German botanist Theodor Hartig suggested raising and releasing parasites from caterpillars. Predators were released in Italy to control the gypsy moth.

Meanwhile, American farmers were expanding into new lands and often the newly planted fields had few pests. Eventually, harmful insect species came to be introduced in the fields, often without predator or parasitic species that tended to control the pests in their native environment. This frequently led to sudden and massive outbreaks of crop damage by uncontrolled insect pests.

A destructive wheat pest known as the Hessian fly was first seen in America around the time of the Revolutionary War. It was rumored to have been either unintentionally brought over by Hessian mercenary soldiers or perhaps even intentionally to destroy the new republic. More likely the fly hopped a ride amongst materials being brought over from Europe, in straw bedding for horses or among wheat straw forage that was brought over in large quantities for military horses.

However it arrived, the Hessian fly found waiting fields of wheat around New York from where it began its relentless spread. The situation was a shock to local farmers who had not before experienced the sudden introduction of a previously unknown and destructive pest.

Thomas Jefferson and James Madison toured the countryside by carriage in the spring of 1791 surveying the damage throughout New York and Connecticut. Jefferson was a skilled farmer and keen observer, and his detailed notes documented the unfolding crisis. He noted that the Hessian fly first appeared on Long Island around 1776, and then in 1779 began to spread westward from Manhattan to Perth Amboy, New Jersey, northward upriver to Poughkeepsie, and eastward reaching Middletown, Connecticut, by 1786. Several years later the Hessian fly was in northern New York state and Canada.

George Morgan noticed that yellow bearded wheat was much more resistant to the fly, which preferred the commonly planted white wheat. He sent samples to

An 1890 cartoon about the Hessian fly from the British satirical magazine Punch. This fly has plagued wheat farmers, including George Washington and Thomas Jefferson, for hundreds of years. It was one of the pests that Herbert Smith studied in his early career, but it remains a problem to this day.

George Washington, who became an advocate for the new variety. Washington noted "White Wheat must yield the palm to the yellow bearded, which alone, it seems, is able to resist the depredations of that destructive insect." Even though it cost almost twice as much as white wheat, many farmers started planting the yellow bearded variety.

While the involvement of several of the founding fathers in battling the Hessian fly problem is a historical curiosity, of greater importance was the wide recognition among the agricultural communities that education and experimentation were needed to improve their lot. Besides trying different strains of wheat, farmers tried out crop rotation, deep plowing, and fertilizing. What turned out to be the most important improvement was the clearing of fields after harvest of any wheat, grasses and weeds that the Hessian fly needed to lay eggs for the next generation.

One of the key figures in the application of scientific methods to entomological problems in the United States was Charles Valentine Riley. Riley was born in London and was well educated in private schools in France and Germany. He emigrated by himself with few funds at the age of 17, finding work in Illinois as a laborer on a farm belonging to an expatriate family friend. From there Riley moved to Chicago where he became the entomology editor for the well-known agricultural publication *The Prairie Farmer.* After a brief stint in the Union Army during the Civil War, Riley at the age of 24 got an appointment as the first State Entomologist for the State of Missouri.

Charles Valentine Riley was a pioneering entomogist who has been called the father of modern biological control. Riley collected parasitoids on a plum curculio in Kirkwood, Missouri and sent them around Missouri. Riley was the first to ship parasites to others to control a pest.

It was in 1870 in Kirkwood, Missouri, that Riley earned the title of father of modern biological control. The plum curculio is a highly destructive weevil that is capable of damaging a wide variety of fruits. Riley collected and shipped parasitoids of the plum curculio that he identified there to other areas of the state.

In 1873 Riley continued his pioneering work in biological control by shipping the American predatory mite *Tyroglyphus phylloxerae* to France to combat the phylloxera that was damaging grapevines there. For this he was decorated by the French government. The next four years Riley studied a plague of grasshoppers in Western states, which led him to convince the US Congress to establish the US Entomological Commission.

In 1878 Riley married in St. Louis and moved to Washington where he was appointed the US Department of Agriculture Entomologist. In 1883 he was involved in importing the insect parasitoid *Apanteles glomeratus* from England to control cabbageworm in the States. Riley was also appointed insect curator for the Smithsonian Institution in 1885.

Riley's life was cut tragically short in 1895 at the age of 52. While riding a bicycle down a steep hill near his Washington home, his wheel struck a paving block dropped by a wagon. Riley was thrown, fatally fracturing his skull. He left behind a wife and six children. In his brief life Riley authored a remarkable 2,400 publications and brought economic entomology and biological control into the modern era.

❖ ❖ ❖

The United States Congress declared war on Germany on April 6, 1917. Herb Smith enlisted in the Navy and was among the first Americans sailing to Europe to join the conflict. Smith served for two years as a wireless operator on Subchaser 257. The subchasers were small narrow ships designed for locating and destroying German U-boats. The North Atlantic was frigid and the dangers included not only the German Navy but also storms and German mines.

While on leave one day from the subchaser, Herb met a pretty and vivacious Scottish girl named Catherine McMillan Wilson. Catherine lived in Glasgow with her family and worked as a shorthand typist. The war effort took on a whole new dimension. Whenever Smith had the op-

Herbert Smith served on the subchaser SC-257 in the north Atantic during World War I .

Top: SC-257 underway at sea. (Subchasers.org)

Bottom: SC-257 crew. {Smith family collection)

Top left: The North Atlantic winter left
the ship covered with ice.
Top right: SC-257 attacking U-boat with depth charges.
Bottom: A German mine.
(All photos from Smith family collection.)

portunity, he returned to call on the home of John Wilson and family at 159 Shamrock Street, Glasgow.

Armistice was finally declared on November 11, 1918 and Smith joined the cheering throngs at Plymouth harbor. Victory was bittersweet, however, as he realized the Navy would be sending the crew of SC-257 back stateside, away from Shamrock Street and Catherine.

When the troop ship landed back in New York, Smith took a position as an entomologist with the United States Department of Agriculture. His desire to see Catherine however led him to soon leave that position for a while and to sign on with the Merchant Marine. His wartime

Herb Smith was in this cheering crowd at Plymouth, England as Armistice was declared in 1918 (Smith family collection).

experience as a radio operator meant he could easily find work on transatlantic freighters. For two years he sailed the Atlantic and Mediterranean, stopping at Glasgow to visit Catherine whenever he could.

One of the ships that Smith crewed on was the S.S. Muskogee. A 7,000 ton steam-powered oil tanker, the Muskogee had plied the Atlantic since 1913. Smith had no problems in his time on the ship, but years later in March of 1942 the Muskogee disappeared without a trace, with the fate of its 34 crew unknown. Decades later, the captain's son George Betts, still having no clues about the fate of his father, began an investigation. While researching German records he discovered a German submarine had torpedoed the ship. He got in touch with the U-boat captain Reinhardt Hardegan to learn about the fate of the crew. He found that there had been seven survivors spotted but the Germans had left them behind on the open ocean.

Herb Smith as a Merchant Marine radioman.

Smith working as a radioman on the oil tanker S. S. Muskogee, c. 1920. Years later in the early days of World War II, the ship was sunk by a German U-boat with the loss of all hands (Smith family collection).

Herbert Smith with colleagues at the Department of Agriculture Insect Lab in Carlisle, Pennsylvania, 1920s. Smith started out his entomology career there researching natural enemies of the Hessian Fly and the European Corn Borer among other insect pests. (Smith family collection).

In 1920 Smith rejoined the Bureau of Entomology and Plant Quarantine, U. S. Department of Agriculture, and was assigned to the Cereal and Forage Insect Investigation Laboratory in Carlisle, Pennsylvania. Along with fellow entomologists P. R. Myers and C. C. Hill, Herb primarily researched the wheat pest known as the Hessian fly, the same one that had bedeviled early planters including Thomas Jefferson and George Washington. Smith was involved in significant investigations at the lab, and traveled widely to survey wheat farms around the Atlantic seaboard. He published several papers on Hessian fly and wheat stem sawfly parasites. The lab also did research on new pest, the European corn borer that arrived in the U.S. from Europe in 1917.

Herb now had a steady job, but he still missed Catherine. They decided to marry and on September 10, 1921 Catherine arrived in New York on the liner S.S. Cameronia on a fiancé visa. They married in 1922 at Manhattan's "Little Church Around the Corner" and soon the pair settled into life in the small town of Carlisle. There they had two sons who sadly died at birth. But they were blessed with two daughters who thrived, Janet and Patricia. Herb Smith now had a growing family and career, but he had dreams beyond small town America, and there was a world out there that beckoned him.

3

France: Life on the Riviera

The European Parasite Laboratory was a US Department of Agriculture outpost in France that primarily collected parasites that lived on European pests. The hope was that some of these parasites could help control insect pests that had made their way to America. The Canadian entomologist William R. Thompson joined the US Department of Agriculture and was sent to France in 1919 to establish a laboratory to study parasites of the gypsy and brown-tail moths and the European corn borer. Thompson set up a lab in the south-central French town of Auch.

The European corn borer (*Pyrausta nubilalis,* also known as *Ostrinia nubilalis*) became a particular focus of the lab. This pest, obviously native to Europe, had only recently arrived in the United States. Maize is not native to Europe, and the corn in the insect's name refers to the older broader definition. The pest originally affected millets, sorghum and a sorghum variety called broom corn that

In 1928 Herbert Smith joined the European Parasite Laboratory in Hyères, France. The town had long been a popular vacation spot on the Côte d'Azur, the French Riviera.

Until moving to Paris in 1936, the European Parasite Laboratory was located in a converted villa on the Avenue Alexis Godillot in Hyères (Smith family collection).

The chief of station Harry Parker with his wife Harriett. Parker came over to assist lab founder Thompson in 1920. He remained in France for two decades. During the war years Parker served in Montevideo, Uruguay, returning to Paris after the war.

was used to make brooms. Once in the Americas, however, it did find maize to be quite to its liking. As the corn borer spread quickly throughout the eastern U. S., the European parasite laboratory became an important outpost to locate and import natural enemies of this pest.

The field of biological control of insects was growing. There had been a number of great successes using natural enemies against harmful insects. The insecticide field on the other hand was still in a primitive state. As the twentieth century evolved, however, the chemical control industry became large and powerful, and the biological control field faced increasing difficulties. From a commercial perspective, it is difficult to profit from parasites and predator insects in the way that business can from the sale of agricultural chemicals.

After a short while the American lab relocated to the picturesque Riviera town of Hyères. The coastal town became popular among British aristocrats after the visit of the Prince of Wales in 1788. Joseph Conrad and Robert Louis Stevenson lived and wrote in Hyères for some years. Even Queen Victoria stayed for three weeks in 1892. Some Americans came as well. For the last twenty years of her life writer Edith Wharton spent every winter in Hyères.

The lab was a converted residence on the Avenue Alexis Godillot. In January 1922, a Bureau of Entomology scientist named Harry L. Parker sailed to join Thompson in Hyères. After Thompson retired, Parker took over as chief of the lab.

In 1928, Herb Smith jumped at the chance to join the group there. He had enjoyed the time he had spent in Europe in the Navy and Merchant Marine, and the move brought his family with their two small children closer to his wife's relatives in Glasgow.

Other scientists from the US came to the lab for shorter periods to do European research. Kenneth Babcock spent three years in Europe studying the European corn borer, returning to the US in 1927. Arlo Vance then followed him to continue that work in the late 1920s. Gilbert J. Haeussler came in the early 1930s to study parasites of

Herbert Smith in northern Italy with Italian entomologist Scognamiglio, 1930s. One summer Smith took his family on holiday in Lombardy while he worked in the fields studying the European corn borer there (Smith family collection).

the oriental fruit moth. In the late 30s Wendell F. Sellers joined Parker and Smith at the European lab.

In addition to the European Parasite Laboratory, the Division of Foreign Parasite Introductions also oversaw a laboratory housed in the Canadian Pacific Building in Yokohama, Japan. The lab led by R. W. Burrell studied the natural enemies of the oriental fruit moth, Japanese and Asiatic beetles, the pink bollworm, and the spruce sawfly.

Both of these labs studied native parasites in the field, but they also collected them for shipment to researchers both in the Department of Agriculture as well as to states and other countries.

Smith traveled extensively throughout Europe, and the European corn borer was the major concern. Its life cycle was studied in Hungary, Yugoslavia, Czechoslovakia, Italy and France. Smith also researched across North Africa from Morocco to Egypt.

From the fall of 1928 to the end of 1930, Smith was primarily occupied by studies of a natural enemy of the

Smith and his colleagues would often purchase agricultural produce to examine and collect parasites. In this photo Scognamiglio is shown with alfalfa stems being prepared for shipment to California (Smith family collection).

Herb at work studying parasites with a microscope. He completed detailed studies of several important parasites while based in Europe.

European corn borer (*Pyrausta nubilalis*). This was an ichneumid parasite called *Phaeogenes nigidens* Wesmael. *Phaeogenes nigidens* was at the time the only parasite of any importance that was able to attack the pupal stage of the corn borer.

While major insect pests usually have common names, their parasites more often than not had none. While farmers tended to be well aware of the insects harming their crops, parasites that attacked the pests were so small that a farmer could hardly notice their presence, much less identify them.

Smith traveled widely to study this parasite in the field. This was necessary as the habits of insects can vary

Catherine Smith with her daughters Janet and Patricia, Hyères, France, early 1930s (Smith family collection).

in different climatic and geographical zones. Smith found seasonal differences in different locales such as central France around Tours, the Jura Mountains, southwest France, and the Mediterranean region around Hyères, as well as areas around Turin, Bergamo, Placenza, Padua, Treviso, Udine and Hungary.

Interestingly Herb found the parasite only had limited effectiveness in much of Europe except in one zone,

Life at home and the beach in Hyères (Smith family collection).

the area around Bergamo, Italy. As a result he spent a lot of summers researching in this area.

Another insect of concern was the alfalfa weevil (*Hypera postica*), one of the most damaging pests of alfalfa. It was accidentally introduced into the U. S. near Salt Lake City in 1904. The scientists at the European lab found several interesting parasites of the weevil. Smith studied one parasite called *Dibrachoides dynastes* in detail, and published a paper on the insect in 1930.

As time permitted, Smith worked on a monograph describing his research with the European corn borer par-

Entomologists Gilbert Haeussler, Harry Parker and Herbert Smith climbing in the Italian Alps, 1931. Haeussler did field work in Japan after leaving Hyères and later became a Department of Agriculture administrator in Washington. In the war years Parker ran a departmental lab in Montvideo, Uruguay and returned to Paris after the war (Smith family collection).

The entomologist mountaineers were amateur climbers but nevertheless managed to safely summit a number of peaks in the Italian alps. They found the climbs were a relaxing break from the long hours of intense field work. They sang drinking songs back at the mountain inns, accompanied by Parker on a harmonica and Smith on his concertina (Smith family collection).

asite. The detailed work, almost 50 pages in length, was published as a technical bulletin by the Department of Agriculture in 1932.

During the years 1919 to 1940 over 23 million larvae of corn borer parasites were collected by the European Parasite Laboratory and imported to the US, and another 3 million were brought from Asia. From these, millions of adult

Herb and Catherine Smith visit the beach at Nice, France with their two daughters, early 1930s (Smith family collection).

parasites were reared out. Other parasites were collected and imported in the cocoon or pupal stages. Twenty two species of parasites were extensively colonized in areas of the United States infested with the corn borer. Two and a half million imported parasites of various types were released. An additional four million parasites were obtained from lab or field collections of future offspring of these immigrant parasites.

(Top) Smith with Harold Powell in Meknes, Morocco, 1930s. Powell was an English pharmacist and amateur entomologist who owned the English Pharmacy in Hyères, France and later a pharmacy in Meknes.

(Bottom) An elaborate pigeon house visited by Smith in North Africa.

Smith worked with the Egyptian entomologist Kamal and his staff at their lab and insectary at Giza, Egypt, 1930s. (Smith family collection).

(Right) Smith and Kamal.

(Bottom) The Egyptian entomology staff.

Herbert Smith with an unknown colleague visiting the Great Pyramids in Egypt, 1930s (Smith family collection),

The insectary at Giza, Egypt, 1930s (Smith family collection).

Catherine Smith, Hyères, France, 1930s
(Smith family collection).

In May 1934 the Department of Agriculture reorganized its hierarchies and created a new Division of Foreign Parasite Introductions. The division was intended to bring together all of the foreign parasite work into one administrative group that would oversee all collections and clearance through quarantines. At the same time it kept most of the department's parasite experts away from the mainland divisions dominated by what Smith called "insecticide men." Any decision to bring parasites into the country for release would be made by other than those who studied the parasites. The bottom line was that if Smith was going to continue in this field, he would be working abroad for most of his career.

Smith and his family found life in the south of France to be idyllic. When there was free time, he enjoyed tennis, golfing and bowling, along with days at the beach and travelling.

Herb often went with men from the lab on mountain climbing and trail-hiking sojourns in the Maritime Alps, the Bergamasque Prealps, and the Dolomites of the Alto Adige. One of the peaks they summited was the 8,271 foot Pizzo della Presolana in Lombardy. The troupe also completed the climb of the even taller Monte Gleno (9,455 feet), passing on the way the ruins of the Gleno Dam. The dam, completed in January 1923, had collapsed less than a year later due to poor construction, killing 356 people in the valley below.

Herb not only traveled extensively around Europe, he also frequently worked in North Africa from Morocco across to Egypt. In Morocco he worked with an English pharmacist and amateur entomologist, Harold Powell, who had run the English Pharmacy in Hyères and later a pharmacy in Meknes, Morocco. Powell would collect Moroccan insects for various entomologists. In Egypt Smith would collaborate with the entomologist Kamal, who ran a laboratory and insectary in Giza.

Occasionally when the children were out of school the family would accompany Herb on field trips around Europe. One summer they all went to Bergamo, Italy.

Herb and his daughter Janet on board the S. S. City of Baltimore returning for a visit to America, 1937 (Smith family collection).

Herbert and Jean Smith, late 1930s (Smith family collection).

In 1936 the European Parasite Laboratory moved to a converted villa on the Avenue Alfred Belmontet in Saint-Cloud in the western suburbs of Paris. The staff and their families enjoyed life in the "City of Light" but the increased living expenses and a poor exchange rate created difficulties for the American scientists on government salaries.

On August 31, 1934 the bright and gay life of the Smith family in Hyères turned dark and tragic. Herb's wife Catherine died suddenly in childbirth along with their new-born son. The family grieving for their sad loss had practical problems to deal with as well. Herb's two young children had lost their beloved mother and his

While in Paris, Smith's family grew with the birth of his third daughter, Lucienne (Smith family collection).

work required him to be out in the field most of the time with frequent foreign travel.

Smith considered an offer from his brother Lee to take in Herb's daughters at his home in Ohio, but he decided instead the children should move in with their maternal grandparents and their Aunt Jean in Glasgow, Scotland. That way Herb could more easily get over to visit with his children as often as possible.

After a time Smith invited Jean to move with the children back to France. In 1935 Herb and Jean decided to marry and raise the two girls together, and Herb was happy to be reunited with his children.

The detailed official reports submitted by Herb to his department back in Washington makes clear that the work was intense and time-consuming but productive. In 1936, for example, Herb was busy collecting parasites of the Mediterranean fruit fly (*Coratitis capitata*) from North Africa for shipment to researchers in Hawaii. That fall 85,734 corn borer larvae were collected of which 5,454 contained larva of the corn borer parasite *Pyrausta nubilalis.*

In October 1936, the European Parasite Laboratory made a major relocation to the suburbs of Paris. The reasons given for the move was that there were increasing research needs in northern Europe and a desire to bring the scientists closer to other research facilities and transportation links. The group set up shop in a converted eight-room hillside villa at 91, Avenue Alfred Belmontet, St. Cloud, overlooking the west bank of the Seine across from the Hippodrome de Longchamp.

The lab personnel consisted of Parker, Smith and Sellers, along with two lab and office assistants and a field helper in Italy. One drawback was a major increase in expenses for the lab and its personnel as a result of higher costs and the devaluation of the French franc. There was a small increase in quarters allowance but that only partly offset the situation. In his annual report Parker predicted the staff with children might elect to return to the states unless compensation was substantially increased, as Parisian expenses were too much to bear.

In 1937 Herb Smith continued to work on the European corn borer and the Hessian fly. He was also studying parasites of some harmful beetles of the genus *Bruchus,* commonly known as bean weevils. Smith worked with Roger Almoch as Hyères field agent to organize groups of workers to help collect the *Bruchus* parasites. Augusto Merselli helped in a similar fashion in Poggio Rusco, Italy for European corn borer collections.

At the end of 1937 Herb took his accrued leave and the family sailed on the liner S. S. City of Baltimore for a Christmas visit to the United States. After this break, the Smiths returned to Paris, hopeful that their lives there could continue as before, notwithstanding the growing talk of war on the continent.

By 1939, Parker and Smith were the only Americans at the lab, not counting P.A. Berry who was mostly in Scandinavia working with the spruce sawfly. They were assiste by local staff — Genevieve Longhais, Jacques Balakine, Otto Conde, Serge Konchalevitch, Raymond Vassal, Lucelle Gailhaonst, and Mr. Chevtchouk.

Herb was mainly working on the Hessian Fly as well as a new project with a pest known as the pea moth (*Cydia nigricana*). Monsieur Balakine helped him with the work. In May Smith made arrangements with local agricultural agents and pea canning factories in the west of France. He got assistance on the pea work from Dr. Tromelot at the Central Experiment Station at Versailles. Herb also arranged with Dr. Taranto in Italy to do parallel studies on the pea moth there. In June a new insectarium was finished and went into operation on the Chemin de l'Empereur in Reuil Malmaison.

On September 1, 1939, Germany invaded Poland. France and England declared war on Germany two days later. In October the American Bureau Chief notified the staff of the European Parasite Lab that the facility would be closed and the scientists and their families would be returning stateside.

Thirty cases of items from the lab including notes, microscopes, and book were stored in the cellar of the US Embassy. Insect related equipment was left at the new insectarium at Reuil Malmaison. Jacques Belakine, a long-time employee, was to remain on the payroll as guardian of the insectarium, for which he would be paid 15 francs per day.

The US government Ford used by the station happened to be in Norway, where P. A. Berry was working on a project. The car was sold there and Berry caught a ship on October 16th heading west from Bergen.

The pea moth cocoons that Smith was working on in the insectarium were shipped to the lab at Moorestown, New Jersey. Even though the trip took a month, the pea moths arrived in good condition.

Herb's wife Jean and the three children were evacuated to Dinard on the north coast of France. Later the family was reunited with Herb at the port of La Verdon in the west of France.

Regular passenger service ended with the outbreak of war. The Smith family caught one of the last ocean liners sent to evacuate Americans from France to the US on October 26th. The S.S. Washington was one of the best American liners, but for this trip the ship was overloaded with evacuees, some of whom had to sleep in the empty swimming pool. The passage was tense due to the risk of

Formal portrait of entomologist Herbert D. Smith (Smith family collection).

attack by marauding German U-boats. Just weeks earlier on September 3rd, a few hours after Britain declared war on Germany, the S.S. Athenia was torpedoed by the German submarine U-30, resulting in the deaths of 118 passengers and crew. Fortunately the Washington made safe harbor, steaming into New York on the first of November.

The S.S. Washington was called up one last time in the summer of 1940 to pick up American stragglers who had not yet made it out of France, as Nazi troops made their way to the outskirts of Paris. The ship called at the neutral port of Lisbon, Portugal, bound for Galway, Ireland. The ship captain, Harry Manning, had been in the news a few years earlier as a navigator on Amelia Earhart's first attempt to fly around the world. Lucky for him his duties prevented him from traveling on her tragic second attempt.

Manning notified all authorities of his route and of the Washington's status as a neutral ship. Nevertheless, on June 11th, a blinker message was received in the predawn darkness: "Stop, heave to." Manning complied but the next message was "torpedo the ship" with a notice they had ten minutes to abandon ship. A general alarm was sounded and the passengers began an orderly transfer into the life boats. Meanwhile the Washington blinkered back the message "American Ship" over and over. After ten agonizing minutes a reply was received "Thought you were another ship, please go on."

There was uncertainty all over as the world was gripped by danger and fear. Europe and Asia were already at war and America soon would be as well. After a decade on the continent, Herb and his family had returned to America and a new life.

4

US: The Home Front

Back in America, Smith set up shop at the Department of Agriculture laboratory in Moorestown, New Jersey. After a while he transferred to the field laboratories in Hoboken, just across the Hudson River from Manhattan. Entomologist Ted Gardner worked alongside Herb. Their group was still part of the Division of Foreign Parasite Introduction, though their evacuation stateside made their futures uncertain given their lack of authority to introduce foreign parasites into the United States.

Stateside insect control was the province of the insecticide men. Robert Burrell and John Maple were for the moment still doing foreign parasite studies in Yokohama, while Harry Parker and Paul Berry relocated and set up a new lab in Montevideo, Uruguay. Back in Hoboken, however, there wasn't much to do except raising parasites in the lab. Smith hated the assignment. He was used to much more active work.

One positive aspect of the posting was that Smith was able to acquire a nice home for his family in Rutherford, New Jersey, a leafy suburb with a small town feel that was still just a short commute from the big city. It was a place where his children could have a more stable upbringing and not have to move around and change schools endlessly. But in spite of approaching middle age, it turned out Smith still had a long career of foreign field work ahead of him.

5

Greece: Picking Up the Pieces

Smith's new life in America involved rather mundane tasks at Hoboken raising parasites and inspecting infested materials. This was interrupted by his blackfly work in Mexico in 1943, and again in 1944 when he accepted a two-year assignment with UNRRA, the United Nations Relief and Rehabilitation Agency. His colleague Ted Gardner had recently accepted a commission in the Armed Forces, but Herb with three children to support and nearing fifty years of age was not what the military was looking for. He had done his part anyway a quarter of a century earlier on a subchaser in the North Atlantic prowling for German U-boats. But with UNRRA Smith could help in the war effort.

UNRRA was formed in 1943 and predates the organization known as the United Nations. Allied planners estimated that 22 million Europeans were displaced from their homes by the war, half of whom were outside of their home country. UNRRA was largely funded by the US and to a lesser extent Britain and Canada, and was created to "plan, co-ordinate, administer or arrange for the administration of measures for the relief of victims of war in any area under the control of any of the United Nations through the provision of food, fuel, clothing, shelter and other basic necessities, medical and other essential services." As formerly occupied areas were liberated, there was a humanitarian crisis among the millions of people who had been displaced and were suffering shortages of food, shelter and medical attention. The program spent three and a half billion dollars over six years, mostly on food and other supplies, but that was swamped by the estimated $350 billion in damage caused by the war.

In Cairo in the summer of 1944, plans were being drawn up for a program of relief to follow the coming assault to remove the German army from Greece. On September 30, 1944 the first group of UNRRA staff arrived on the island of Chios. The rest of the Greece Mission meanwhile gathered at a training camp at Maadi, Egypt. In the next few months most of the UNRRA personnel made their way to Greece and began operations. Many villages had been burned by the Germans and many inhabitants had been killed. Care of destitute children became a pressing focus.

The UNRRA effort in Greece over several years brought over $400 million in supplies, half of it food. The situation was complicated however by Greek government interference and local corruption. Black market profiteers waylaid some of the goods, and much of the food and supplies never made it out of Greek warehouses. In 1947 an American inspection team discovered mountains of American textiles sitting in warehouses, having been diverted to protect Greek textile manufacturers. They found American UNRRA foods on the black market while three fourths of Greek children were still suffering from malnutrition.

Although these problems were beyond Smith's control, he did his best to help. He worked in Greece through the end of 1945, returning to New York on the Liberty freighter S.S. Hannah S. Abell in December. There he was reunited with his family and began work again at the Hoboken Lab.

6

Mexico: An Industry in Peril

Smith first took up the citrus blackfly problem in 1943. Unfortunately the parasites that Herb and his Mexican counterparts imported and released around the west coast of Mexico did not prove to be as effective as they had in the Caribbean and Central America. The Pacific coast had a far different climate. While the parasite did well in the warm and humid Caribbean, it didn't fare well along Mexican coast with its dry and cooler winters.

The citrus blackfly continued to spread around Mexico, and in June 1947 an infestation of citrus blackfly was found at Empalme. Since this was only 260 miles from the U. S. border and along a key rail line from Nogales, Arizona, the level of concern among the American citrus industry increased markedly.

Herb had kept an interest in the blackfly problem and was disappointed that the parasites he had brought to Mexico were not doing the job. In 1945 he published a lengthy scientific paper describing the 1943 effort there. Interestingly Smith submitted the work in Spanish to the Mexican journal *Fitofilo,* which had only recently been launched to cover biological control of insects. Alonso Blackaller of the Oficina Fitosantaria helped with the translation. Smith had realized there was a strong interest in biological control in Mexico, greater than in the US which had no similar journal. In Mexico Smith was finding an attentive audience and a growing group of scientists with interests similar to his own.

In 1947 Smith undertook a second attempt to deal with the worsening Mexican blackfly problem. It was thought that introducing a second parasite or predator species into Mexico might work. It had been reported in Cuba in 1931 that a species of predator beetle known as *Catana clauseni* aggressively attacked the citrus blackfly in the few citrus groves into which it had been released.

A few years later Catana was no longer found on the blackfly, which itself had been reduced to low numbers. Dr. Bruner had seen Catana feeding on Cuban bamboo scale, however, and in 1938 the entomologist P. A. Berry collected hundreds of them feeding on whiteflies of papaya. Smith contacted Bruner again who reported that Catana had not been noticed since 1938, but that no one had been particularly looking for it. Smith thought since Catana had survived in Cuba for at least seven years after release, he might still be able to find some there.

Herb left New York on October 6th, traveling to Washington to consult with various specialists and officials. The next day, he boarded a southbound flight, arriving in Havana two days later. Needing local transportation, Smith met with US consular officials who were only able to offer a rickety old army truck, and then with Dr. Brunner of the Cuban Department of Agriculture, who managed to scrounge up a jeep. As was his usual practice, Smith made the rounds of courtesy visits to varied local officials - in this case the Assistant Minister of Agriculture Grau, the Directors of the Agricultural Experimental Stations, Dr. Portuando of the Ministry of Agriculture, and Jorge Des Chapelles, Chief of the Seccion de Sanidad Vegetal.

Smith then got down to business scouting in the field, accompanied by Dr. Brunner's assistant F. Valdes Barry. They were searching for Catana, which had been introduced into fields in Havana province at two locations, one near the Havana airport at Santiago de las Vegas and on the road to Mariel at the village of Caimito. Thorough searches were conducted six days a week for more than three weeks, but Catana was nowhere to be found.

The citrus trees there had only a few scattered blackfly clusters, but the papaya trees had heavy whitefly infestations. Bamboo in the area, which had also been infested with Catana in earlier surveys, as well as other plants that might be infested with insects of interest to Catana predators, were all carefully investigated. Smith eventually scouted "all passable roads in Havana province." Groves, backyards, botanical gardens, and nurseries were also covered. The search was extended to Pinae del Rio and Natanzas provinces to no avail. Catana had done a disappearing act.

Smith carried nine Wardian cages weighing a total of one and a half tons throughout his travels. The cages contained live plants and the pest insect species, all needed to bring back live parasites. He discovered that international air routes newly available after the war provided an easier and more reliable way to import parasites, as they could often survive a week or more in an air express box (Smith family collection).

❖ ❖ ❖

While the citrus blackfly was aggressively spreading through the citrus groves of Mexico, its behavior in its Asian homeland was far different. With the exception of occasional flare ups here and there, the citrus blackfly was in general a minor pest there. It seemed likely some native parasites or predator insects were keeping things under control. As a result a decision was made to launch an effort to recover different controlling species and import them into Mexico.

On March 16, 1948, the U. S. Congress voted to appropriate three million dollars for insect investigations, with part of that earmarked for dealing with the blackfly. P. N. Annand, chief of the Bureau of Entomology and Plant Quarantine told Congress that unless the blackfly was stopped in Mexico, American citrus groves would be in peril. He said that the department entomologists would try to find biological controls from the blackfly's native Asian habitat, and if that failed then chemical means would be used to attack the pest.

With funding in hand, the department wanted someone to do an extensive search for parasites of the citrus blackfly in Asia. Herb was by this time well into his fifties with dependent children at home. No one would have blamed him if he refused the task, but he was the most qualified person for the job, with several decades of experience with insect parasites.

Some of the other government entomologists had done extensive field work, but that was in their younger years for the most part. Clausen had spent time in Japan, Malaya, and India among other places. His Hoboken coworker Ted Gardner and Haeussler, his colleague from the American lab in Hyères, had both worked in the field in Japan and elsewhere. The younger department entomologists mainly went into insecticide work. So Smith volunteered for the task.

Smith prepared for extended field research on the Malay Peninsula that was expected to last many months. He would start there because that is where Clausen found the parasite that worked on the citrus blackfly in the Caribbean. In order to collect parasites in Malaya, Smith would need to take with him blackfly infested plant material to sustain the parasites on the trip back.

In May Smith traveled to Cuba, staying at the Ocean Hotel on Havana's broad seafront esplanade known as the Malecon. He was extremely busy getting ready for his Asian trip. He had no car there, so he took lots of buses and walked a lot.

While in Cuba, he was trying to arrange passage to Singapore but wasn't having any luck. There was no government agent in Havana to arrange things and he was getting no replies from his letters to steamship companies. Most of the ships seemed to be booked up.

He wrote to his daughter Janet in Paris to see if she could find a ship that could get him and his cages to Singapore in June or July. Janet consulted with her French in-laws, who were familiar with international shipping because of the family import/export business. Smith was willing to go via New York and the Atlantic or through the Panama Canal and across the Pacific. In early June Smith was still in Havana with no passage to Singapore.

Most of July was spent in the Canal Zone in Panama, but the blackfly was so well controlled there by parasites that few could be found. Most Americans tended to stay in the Canal Zone, but Smith had to travel to outlying areas in his search.

"They had a revolution the other night in Panama, practically across the street from this hotel," he wrote home to his family. "I was out walking around in the city just a couple of hours before it happened. About 35 injured and five dead. Not much trouble since except isolated killings." The Canal Zone itself presented a far different atmosphere. "This is like living in the U.S. They have clubhouses for the people. Many things are cheaper here than at home. You get free doctor care..."

Smith wired Clausen to try to get some infested material from Mexico. Eventually he received four cartons of infested Mexican leaves. "The bugs are boiling out. Clausen should have arranged this long ago, and I'd have been in Malaya by now." Smith reflected on the uncertainties and serendipity involved in doing field work. Clausen had done the field work years earlier that brought back the parasite that controlled the blackfly in the Caribbean.

> *Clausen made a name for himself on this job. By luck the parasite he introduced was not the one he intended, but it took hold and wiped out the pest. The ones he was to introduce he failed on and that is now my job. Of course it wasn't his fault, but it just goes to show what a great part luck plays in this game. So far I haven't had any luck.*
>
> *I knew all along that getting these bugs from Mexico was the thing to do and told Clausen so. He thought I should try Panama and so I did. The Mexi-*

can leaves came through without any parasites, thank goodness. If that parasite had been in them I would have had an awful time as it would have killed all my Cuban stuff. I'd have had to sift them out, a terrible job. The parasite is the size of a pin point…you don't see them unless they move, and they don't move much.

As no passage on westbound ships was available for weeks, Smith instead sailed eastward. He embarked July 23rd on the Panama Line steamer S.S. Panama with one trunk and many large cages, arriving in New York harbor five days later. He had booked an ongoing passenger liner for July 31st, but there was no cargo space for his huge cages. Instead he booked passage on the next available cargo ship, a converted World War II Liberty vessel called the S.S. Steel Scientist, now on its third voyage for the Isthmian Lines.

The ship pulled out of New York harbor the afternoon of August 19th, and Smith surveyed his new home for the next few weeks. He had a three-person cabin all to himself, giving him space for his equipment and trunk. The cages were stowed on deck, and the ship's officers and crew were helpful.

There is a nice crowd on board and they are all talking to me while I work on my cages. They all have to see what it is all about. I'm glad I have the plants. It gives me something to do. Every morning I go down and open them up to look them over and scrape off the bugs I don't want. I took out all the moss packing as

> *it harbored snails and millipedes. I found three little frogs in one cage.*

Smith and the other passengers played poker to pass the time. "I am way ahead so far. Play with matches! Maybe we will cash in later. The captain serves us Teacher's Whiskey, and is very generous with it." The weather was sunny yet cool across the Atlantic, and began to warm as the ship approached Egypt.

The ship arrived at Port Said on the 30th, where a couple of oil men and their families disembarked. A young Chinese man, in the states to attend Columbia University and then to work at the Ford plant in Flint, remained on board with his wife. The man's family owned the Ford dealership in Singapore, and he offered to help Smith get settled in nearby Kuala Lumpur. The ship was laid up for a bit in Port Said to make a few repairs and waiting their turn through the Suez Canal.

> *The ships can only go one way at a time. Now we are waiting for the northbound ships to get through… then it's our turn, probably around 2 a.m. It takes 8 to 9 hours through the canal. There is a big searchlight on the bow for the trip.*
>
> *I asked the captain if I could go ashore but "no go." Not even for a quick beer. This captain is quite strict. Too much so…Of course he went ashore with the two families who left to continue by air…One second mate just quit here because the captain was always yelling and bawling him out.*

He criticized me because I wore a T-shirt into lunch one day. We have to wear a shirt or jacket, no tie. The officers have to dress that way so we have to... This captain is a Prussian German type...He caught me on the bridge one day and politely said I was not to be up there.

A few days later the ship anchored off Jedda in Saudi Arabia. The weather had turned very hot. Smith used an awning to cover his cages. The scheduled one-day stop there turned into four. Anchored six miles from shore, it was a constant 95°F. in the cabins. The crew slept on deck where it was a bit cooler at night. The water temperature was 92 degrees and the engine room was 126 degrees.

When the Arabs came on board yesterday they arrived at 6 am and before I could get on deck they had moved my cages off the hatch. They stole my sucking apparatus, some cloths, etc...One thing they didn't take was the flit gun. They don't kill flies, it's against their religion. The flies are so thick ashore that if you open your mouth to talk they jump in!

The whole coast of Arabia is filled with rocky shoals...The Arab fishermen go out in boats and wade around on them for fish or shells.

It was Pilgrimage season for Mecca and several boats came in with Arabs and they got priority...I watched the Pilgrims leave the ships all around us, with the glasses. They carry white umbrellas...The women of course were veiled. The boats are all sail-

boats, like a large row boat, yet they carried our biggest trucks.

The ship pulled into Aden harbor for part of a day, and then proceeded into the Indian Ocean. Herb spent some of the long hours talking with the officers and other men. He avoided the captain, who wasn't very friendly and was somewhat peculiar in Smith's eyes.

They all come around when I'm at the plants to talk to me. None really understands what I'm doing, although I've explained it many times.

The plants are coming along okay and are sending out new shoots. The flies are out and laying eggs. I have no awning over them now (too windy) so they get full impact of the sun. It was 100 degrees in the cages. The sun is hotter here than in the Red Sea... nearer the Equator. I had to quit working in the sun as I almost got sun stroke through my hat and clothes. The heat even burned through my shoes. Now I only work early in A.M. and late in P.M. I've been working in my shorts with no shirt on and have gotten very brown. But brown as I am I still get burned from this hot sun.

Arriving off the Malayan coast, the ship stopped at Penang harbor for a few hours on September 17th, Smith was able to go ashore for the first time in weeks. The first thing he noted was the many English soldiers on the streets in this British colony. "I bought a few beers but the

English beer is so terrible I'd stop drinking beer if that is all I could get!" He bought a bottle of Canadian whiskey. The ship continued on, the weather turning noticeably cooler, then rainy with rough seas. The wife of the Singapore car dealer got seasick.

> *Not much has happened on the ship. We are all fed up with the captain. He is an old sour-puss and thinks he can dish it out and not take it. The Chinese couple and I don't bother with him at all. He told them he'd like to throw my trees overboard. He refused to put up the awning at Aden. He told me to do it myself. Some sailors helped me in their time-off. Believe me the next ship I go on I'll check the captain first.*

7

Malaya:
The $50,000 Parasite Hunt

The Steel Scientist landed at Port Swettenham the afternoon of the 18th.. Smith had earlier alerted the Malaya Director of Agriculture by wire from Penang. He was surprised to be met by two vice consuls from the American Embassy in Kuala Lumpur, who had driven down in the consular car on a Saturday. Smith noted that one of the staffers carried a revolver in a holster while the other had an old army rifle in the car.

Malaya was in the early stages of what was to become a decades-long Communist insurgency. A few years earlier Western propagandists were keen on pointing out the imperialist ambitions of a militarized Japan. Many citizens of Asian lands colonized by Britain, including India, Burma, and Hong Kong as well as Malaya, took this to mean Britain would give up its colonial empire. They were none too pleased to find the earlier rhetoric had been rather hollow, and the Western powers intended to carry on as before. In Malaya, an armed anti-colonial guerrilla force inspired by the Communist revolution then sweeping across China found recruits primarily from the Chinese minority population, disenfranchised from political and social status.

While the American Smith was a bystander to that conflict, as a white Westerner he was very much a potential target. He was offered a firearm by the consular staff, which he declined. "No Americans killed yet, but quite a number of British," he wrote home to family. The embassy staffers sped Smith through customs, barely managing to fit his trunk and equipment in the consular car. The cages were carefully unloaded in a rope swing onto the dock, but would need to be left behind for the moment.

The Director of Agriculture reserved a room at the ornate, palatial Station Hotel, an English-style hotel run by a Chinese family. This was fortunate as hotel space was in high demand and in limited supply. He shared a room with an Englishman. Tea was served in bed at 6:45, followed by breakfast in the dining room. Dinner was $1.50, with three or four waiters per table. The $6.50 a day room and board seemed exorbitant to Smith. "I could have done with a simple hotel, but over here you go where the other people go."

At the Station Hotel, Smith ran across an old friend and colleague, Noel L. H. Krauss, an entomologist with the Territory of Hawaii Department of Agriculture. Beside the federal agency Smith worked for, Hawaii and California had the most active American biological control pro-

Herb in a grove of trees examining insect pests and parasites with a hand lens (Smith family collection).

grams, supported by wealthy local agricultural interests. Smith had met Krauss years earlier at Hyères. Krauss had arrived in Malaya in 1948 and had been staying at the Station Hotel for a couple of months. This would be a good connection, as Krauss knew all the ropes on how to deal with local officials and so forth.

On the next day, a Sunday, the Consul met him with a car followed by a truck to go back to the port to retrieve the nine Wardian cages. Smith could have taken the train, but the Consul, who had little to do as few Americans visited the area, insisted on helping out. The cages were then placed under the balcony of the Biological Building at the Malay Agricultural Department located at Lake Gardens, a few miles outside of Kuala Lumpur. Later they were moved to a shaded upstairs balcony just off the Entomological Laboratory. Pans of water placed under the cages kept the ever-present ants way.

Smith began his search with scouting trips around Kuala Lumpur with Robert J. A. W. Lever, the government entomologist. The city had lots of backyard gardens that yielded a few leaves infested with the citrus blackfly. He found the town difficult, as he was used to traveling freely.

> *It is safe in the city, and the city covers a lot of ground. There are lots of big gardens. Everything is Chinese with some Indian. The British just run the government. The rickshaws are slow and expensive. The white people don't use them, I found out, they*

use the taxis, but you can't find them very easily. The permanent British have cars.

There is nothing to do here, except the movies. No cabarets and the dance halls are for the Chinese. The British have their Clubs...After Cuba and Canal Zone this place is a big let-down, sort of depressing. Not even a cafe to go to.

Here it isn't really hot so far, but damp. White people just don't go rambling around here like in other places. Things aren't restricted, but it just isn't done. I can't ride on buses, only for natives. I went to an amusement park and was the only white man there.

Due to overcrowding, Smith was only allowed two weeks at the Station Hotel, and had to move to a boarding house. There was no bath or toilet attached, but he had a room to himself. Tea was at 6:45 A.M. with oranges and biscuits, then breakfast at 7:30 with porridge, fish, eggs, fruit and coffee. "Tiffin" was at 1:00 P.M., followed by 4 o'clock tea, and supper at 8 P.M. with soup, fish, meat, vegetables, dessert, fruit, cheese and demitasse coffee. There were four waiters for a table of two.

The government entomology laboratory was made available to Smith. It was impressive, as good as or better than what he had stateside at the Department of Agriculture. There was a large library attached as well.

The growing insurgency made the survey work difficult and dangerous. One of his best collecting areas was along the Ipoh Road, which was also a known hangout for the guerrilla forces. A mill just off that road was a favorite meeting spot for the leaders of the insurgents. Smith was advised against going more than nine miles down the Ipoh Road, but the best tree he found for collecting adult parasites on leaves was at the nine and a half mile mark. Smith made almost daily excursions to the area, checking first with the Constabulary for updates on the security situation. "Every day a few constables, planters, etc. are killed, although not all in this vicinity." Besides the personal risk, the situation caused local Chinese residents to fear his visits. There were few citrus groves, and those he found had little blackfly trouble.

> *My best results have been obtained from isolated trees of backyards in villages (kampungs), around farmhouses, habitations on the rubber plantations, and tin mines. This requires a stop at every house where a citrus tree can be seen from the road and I often check even though citrus cannot be seen. I always travel with a Malayan assistant. As I go unarmed, I have no trouble in entering the properties, but there is always suspicion that I am checking on terrorist activities. I have been offered military escort (a soldier with a gun in the car) but declined as this would hurt my chances in case I wished to return to the region. Most Europeans travel around the country well-armed....*

Smith covered all the kampungs around Kuala Lumpur and the Botanical Gardens, but what little infestations he found were usually of the wrong species, the orange spiny whitefly (*Aleurocanthus spiniferus*) rather than the blackfly (*A. woglumi*). Herb traveled all the way out to the Central Experimental Station at Serdang. He went out to Klang, stopping at 23 gardens along the way. His letters home filled in a lot of details about his experiences.

> *I went out with a Scot from Aberdeen last night. He eats at my table of four. The others are an English lad and a Dutch boy. They whoop it up now and then on Saturday nights but are of good families. Very good manners. I like the Scot better than the English, they are too stuffy. The English men are always well dressed, shoes shined, etc.*
>
> *I have been working every day; 7 days a week …I have to go to work by taxi and don't like it. It costs about 79¢ one way. Have tried every way to rent a car, but it can't be done. Every foreigner has a car here and they are higher priced than in the US. Gas is 70¢ a gallon and rationed. They have small cars that go 35 miles to a gallon.*
>
> *Everything is Chinese; they seem to have taken over the country. They work all hours and hard. Every country house I stop at is Chinese. They are honest but you have to bargain with them. I never saw such an industrious people. The Malayans are lazy but very good natured. They like an easy life. They are bright enough but living is so easy here that they see no use in exerting themselves. Only the Malayans are in the government service. The British run the government.*
>
> *The Chinese love flowers and their gardens are full of all kinds, in pots and cans and everything. And*

the kids are everywhere, Chinese women do much of the manual labor, build the roads, baskets on their heads with stone, carrying wood, water etc. on a stick balanced on their shoulders. I often wonder what they think about. No mental troubles here. They don't worry that they work like slaves. I am stopping at Chinese houses every day to look at their citrus trees.

The kids come up and help me pick the leaves and jabber away. I say thank you when they give me a leaf, but they don't understand. It is annoying to go around and not be able to understand them. You feel foolish. Not one word do I understand. They are very peaceful. I just walk into their garden without asking permission and start looking at their citrus trees.

They smile and do nothing. I have a bit of trouble with the dogs. They all have dogs and they bark something terrible at me. I always grab a stick and am ready. They say the dogs don't like our smell as we are big meat eaters, and they are not.

I haven't run into any terrorists yet and don't want to. One of my best spots is right on the edge of the jungle where they hang out. It is where I am getting most of my material. The two assistants are a bit jittery about going, as they heard shooting the other day. I didn't. Everyday a planter or policeman is killed. They enjoy killing white people. Most of the white people who go in the country carry a gun as it drives them off if few in number.

Then with the scrub typhus that you pick up from mites that are on the grass it isn't too pleasant.

I oil the tops of my socks and cuffs of my pants with a preparation to ward off these mites. I have to walk through that type of grass each time I look at a citrus tree. The mortality is high for one over 40 and that new serum they discovered won't be available for another month here. The black market took it all as the communists in the jungles are exposed to it all the time. There is no inoculation for it as with the other typhus.

It's hard to remember to put oil on your clothes every time you start out. So many times I don't bother. Anyway the grass is often up to my armpits. The mites are carried by rats and as they run through the grass they jump off to the grass, a special kind called "Lallang" which is everywhere. The mites crawl up and wait for someone to pass and get onto his clothing. Then they hunt for his skin and start to work. I've felt great so far but if I get that I'll somehow get that serum from the States that cures it.

I have found a fair amount of parasites and released them in my cages. I hope they go to work. I will work in Singapore later. There is not much infestation around here.

I wish Clausen would write. I was supposed to be here a couple of months in the Far East. I haven't had a chance to study the life here yet. The business part is very interesting – all Chinese, they are so busy they do everything.

A few days later Smith mailed home an update about life in Malaya.

> *Well it's almost a week since I've been here and I haven't accomplished anything. It is hard to get around. You share a basin with the other fellows, also the shower is cold water only. There are no springs under my mattress. It is just lying on boards. I have practically no place for my things and only one center light. There is no mosquito netting. Lucky I have the aerosol bombs that Gardner gave me. I spray the room every night. They say they are not the malarial kind, but you can never be sure.*
>
> *It is a big comedown from the hotel I left, I think I'll see if I can't get back and stay there.*
>
> *They put you out after two weeks, that's why I left. My roommate was a nice chap. He snored a bit and kept me awake though.*
>
> *Here you even have to furnish your own coat hangers and everything. A British boarding house.*
>
> *I went to Mr. Lever's house for drinks before supper last night. I didn't know whether it was for dinner as well, luckily I went with another chap. That's the way they do it here. Drinks from 7 – 9 P.M. and then you go your own way.*

By late September Smith was having some success in recovering small numbers of parasites from citrus blackfly infested leaves. "They are as big, or rather as small as (.) that dot," he wrote, "and I have to separate several spe-

Smith's hunt for parasites of the citrus blackfly in Malaya during the so-called "Emergency" was covered by the local press, such as the *Malay Mail* (right) and the *Straits Times* (bottom). Both articles appeared on October 29, 1948. He was traveling in areas where frequent attacks on Europeans made the work dangerous. He was advised to carry a weapon but declined as he didn't want to alarm the local inhabitants as he searched for the blackfly.

PARASITE HUNT IN MALAYA

Entomologist Looking For A Tree

Mr. Herbert D. Smith, an American entomologist of the United States Department of Agriculture, who is on an official parasite hunt in Malaya said that the U.S. Government had set aside a sum of $25,000 (U.S.) for his work. He arrived in mid-September with nine Wardian cases from Cuba and Panama to collect in Malaya parasites which live on citrus black fly liberation in Mexico.

He hopes to complete his "hunt" by next month. So far, he has been successful in collecting two species of the Malayan parasite, the Prospaltella Divergens and the Prospaltella Smithi. He is now in search of a third specie, the Prospaltella Catana. Mr. Smith said that he had been told that this third specie was found some 20 years ago only on one tree in Malaya. "I am looking for that tree," he said.

He declared that he had been considerably handicapped in his work by the present emergency. He explained that his work was such that frequently he had to penetrate into isolated kampongs and remote villages.

Mr. Smith is one of six American entomologists sent out by the U.S. Government to search for parasites. The other five are at present in Africa and Europe.

Mexico, said Mr. Smith, was at present facing a serious problem caused by the citrus black fly which had been responsible for reducing by almost half the citrus crops. This pest was discovered in Mexico shortly before the war but assumed serious proportions in the last four years during which time Mexican orchards were badly affected. To control the citrus black fly (Aleurocanthus Woglumi) he said that he had been sent out by the U.S. Government to find the biological enemy of this pest.

His work, he concluded, had been greatly facilitated by the ready co-operation accorded by the Department of Agriculture, Federation of Malaya.

1 PARASITE: $50,000

From Our Staff Correspondent

KUALA LUMPUR, Thurs.

MR. HERBERT D. SMITH, an American entomologist of the U.S. Department of Agriculture, is looking "for one tree in Malaya" on which he hopes to find a specie of a parasite last noticed 20 years ago.

Mr. Smith is on an official hunt for parasites which live on the citrus black fly.

He has already found two species of the Malayan parasite. The third is the one reported to have been seen 20 years ago only on one tree.

The U.S. Government has set aside $50,000 (Straits) for his work.

cies." He found few citrus trees in the area, just occasional trees in the backyards of local Chinese residents, and only a few leaves per tree infested with the blackfly. The infestations were more widespread in Cuba.

> *I am glad I brought my black fly over with me or I'd never have gotten anywhere. The communists keep me from going out on a number of roads. I check with the police before I start out, as it changes day to day. No Americans killed as yet. They pop out of the jungle as you go by in a car and shoot. The consul offered me his gun but I'd rather pick roads that are fairly safe. In the city it is quite calm and peaceful.*
>
> *I will have to give up the idea of getting to a hotel with running water, nice lights, etc. More English are coming in all the time – officers and planters from the rubber estates.*
>
> *I go to work by taxi. That is the only way a white man travels here. No buses or rickshas --they are only for natives, yet the buses are often only half filled. I wonder what an Englishman would do in Havana?*
>
> *I have been awfully busy as I cannot get any assistant to do much of the work. It's not like in the states. The boy I now have speaks no English and I speak only one word of Malayan (pigie, to go).*
>
> *I am getting used to the boarding house. I was offered a better room but decided to stay here next to the kitchen. There is too much noise in the other place, with guys coming in drunk and so on. They just have partitions to separate the rooms in the other place. A*

great big room and they made four rooms out of it. The wall starts at a foot from the floor and ends 6 feet from the ceiling. Women are in the rooms too. They all use the one bath (seven of them) and everything is in one room.

It is awful in the morning. Some privacy for a woman when all a guy has to do is bend down and look under the wall. I am in a real room, used by the servants before. We are all men in this section so one can shave and the other wash at the same time. The toilet and shower (cold only) are in separate rooms. I take a bath in a wash tub, stand in it and throw the water on me with a pan. Of course you wouldn't stay here a minute, but I overlook the various things.

In October Smith traveled around Seremban in Sembilan State scouting for blackfly infestations. All available roads around Kuala Lumpur were scouted. Smith continued to be advised by local authorities which segments of roads outside of the city might have less risk of running into insurgents. Some roads could be traveled on, but he was told not to stop and scout for citrus trees. Armed escorts were furnished by the police for certain areas, but Smith generally avoided those as they intimidated the small farmers and residents with citrus trees in their yards. One of the best collecting areas was north of the city on a nine mile stretch of the Rawang road. After a few weeks Smith was advised this area was considered dangerous and that he should stop going there.

Sometimes Herb would go out on a certain road one week and then find it prohibited the next week or for the rest of his stay. Several trips were made to Klang and vicinity as this was a good collecting area.

Eventually only light infestations were found, as his extensive collections were removing what colonies were there. The roads to Serdang, Puchong, Ampang and Pahang were scouted numerous times. All kampungs in Kuala Lumpur were scouted regularly with good results. Most of what Smith was finding at this point were from isolated trees in backyards in the kampungs in Kuala Lumpur and villages nearby. Around Sereraban, all the kampungs and most of the villages were visited.

Smith's work engendered local curiosity. In late October The *Malay Mail* and the *Straits Times* newspapers both published articles about his search. He indicated to the reporters that he "had been considerably handicapped by the present emergency," as the civil war there was euphemistically called. Nevertheless he had already found two parasites, *Prospaltella divergens* and *P. smithi*. He was hoping to be finished within the month. The *Straits Times* reporter was particularly interested to hear that the U. S. government had alloted Straights $50,000 (around US $20,000 at the time) for the purpose of locating one insect.

Herb told the reporters he was still looking for a Catana beetle that "was found some twenty years ago only on one tree in Malaya." This was the elusive predator species which had been released into Cuba and for which Smith had made several unsuccessful searches there. "I am looking for that tree," he said.

On October 20th Herb wrote home with a progress report.

I found the parasites, at least two of them, but that wasn't expected to be hard. I haven't found the bug I was to get in Cuba. It came originally from here and was found only on one tree here. I haven't been finding the quantities of parasites that Clausen did. Not yet anyway. I haven't turned up many males of the parasites and that is the important thing.

He lost all his shipment to Cuba because he had no males when he got there. I would hate to leave here and then get to Mexico and find out I could not breed them just like Clausen. He released several thousand in the field and laboratory but nothing came of it.

I've got to solve that problem before I leave or I will be a laughing stock. After all the money that has been spent, to arrive home with nothing. Which would be about it. Clausen was lucky in that he had another parasite that happened to take hold which covered up the other failure.

This all means I'll have to stay here a bit longer than Clausen said. He always makes things look so easy and sets such a short time. There is a parasite man here from Hawaii and he has been here since May and will be here a couple of more months. Clausen seems to be in a hurry but never says anything if you take longer.

You can't hurry this job. I have been working seven days a week and until 6:30 every night to get my work done. No Saturdays or Sundays off at all. If I could stay three or four months I could work decently and have the weekends off. Anyway I will be here as long as I see fit as I want to be darned sure I'm bringing back something, not just going for a useless ride.

I went down the country Monday and Tuesday. Nice city. Slept in a rest house. Better than here. Had a bathroom and took a bath. I met a Scotch/Irish from Belfast, much nicer than the English. Much more sociable and glad to answer questions about the country. He took me to his home for lunch. It is no trouble to entertain here as you have three servants.

Smith found that citrus blackfly (*Aleurocanthus woglumi*) was generally more common than the orange spiny whitefly (*A. spiniferus*). Limes were usually most affected. In general though very little infestation was found in any citrus grove. Groves of several hundred trees at Serdang, Klang and Puchang yielded only 25 to 50 clusters each. None of these had been sprayed with insecticides.

In a typical scouting outing, Smith might examine 50 trees in backyards along a several mile stretch of road and find over half of them entirely uninfested. The rest might have 5 to 25 clusters per tree. The best collections were found on these isolated kampung trees, mainly limes. Typically out of every one thousand backyard trees examined, only ten showed a fair infestation of more than 100 infested leaves per tree.

Herb continued to write home detailed observations from his travels around Malaya.

The British have run this country for a long time, yet there are very few English here – about 60% are Chinese, 35% Malayan and the rest are English and Dutch.

The other day I took a picture of a Chinese boy and girl of six or seven years. They were awfully shy. I gave them candy wrapped in paper. All the older people stopped to look as though it was something funny. A white man here is considered superior and when he does things like that people look. Of course they think I am British. The Chinese children play a lot with marbles, all white. Every store has them. The younger children go about naked and no one pays any attention.

Then there are the Tamils who are Indians (from India). When the British came here years ago they started rubber plantations and dug for tin, but the Malayans wouldn't work for them. Everything grows easily here so people don't have to work hard to make a living. So the British imported Chinese coolies and Indians from southern India (called Tamils) to tap the rubber trees and work the tin mines. The Chinese and Tamils went into business and control it in most of the country. The Tamils have their own language and are dark skinned.

The Chinese are very nice here. They control all the important business. They are also carpenters, mechanics etc. They are very honest, smart, and wealthy. I have been to tea at some of their homes and they live very luxuriously. Years ago they used to give or sell their girls to the Malayans as Chinese only wanted boys. Now with the better class they make no difference between boys and girls. In China girls are still not wanted so you see lots of laborers here are women (from China). They are dressed in black, carrying sand etc. for buildings, or working on roads carrying gravel. They work for less than the men.

The Indians live in one room, sleep on the table (no mattress). They tap the rubber trees. They have to start at 5 AM or earlier. The girls cut a thin strip of bark, careful not to touch the green or cambium layer, and the white sap flows and runs into a cup. If they cut later in the day it won't flow, it stops running about 8 AM. Everyday they cut the bark and gather the sap. The men lug the pails on a long stick over their shoulder. The reason they produce rubber here is that labor is so cheap (90 cents a day) and it is mostly hand labor.

The Malayans are people who came from Sumatra and the other Dutch islands. They are very short, happy, lazy and nice people. They are Mohammedans because the Arabs conquered this country years ago. They have little ambition or energy, the reason being that in the tropics you can eat by picking the fruits off the trees, the temperature is the same the year round, there is no need for heated houses, or much clothing.

Smith also traveled to Singapore and made scouting trips around the kampungs and towns on the Island.

Infested leaves containing pupae were brought to the laboratory for rearing. Smith trained his local assistant Abdulla to help with the collecting. Two other Malay workers from the Department of Agriculture also occasionally helped with the collecting. They would go out into the neighborhoods by bicycle or bus. The Malay government provided very good laboratory facilities. Cages were arranged around the upper balcony of the Entomological Laboratory out of the sun.

The infested leaves were caged in cardboard mailing tubes. More fresh leaves were placed periodically to add humidity as the weather was quite dry most of the time. Several thousand infested leaves were reared and records kept from 130 localities or trees. A total of 13 male and 837 female *Prospaltella divergens* parasites were recovered. Even more *Prospaltella smithi* emerged – 185 males and 1115 females. Usually both *Prospaltella* species were reared from the same lot of infested leaves. There were local differences in relative predominance. *Eretmocerus serius,* the parasite that Clausen introduced into the Caribbean, was found in large numbers only on a single tree with an infestation of orange spiny whiteflies. About 300 *Eretmocerus* were released in the cages.

An additional 1000 female *P. divergens* and 90 female *P. smithi* were collected on leaves in the field and released in the cages. At the first mile on the Serdang road, collections were started November 3 and continued almost every day until December 18. A total of 929 *P. divergens* but only four *P. smithi* were collected. These were mostly collected by Herb's assistant who became quite proficient in this work.

In late November Smith wrote to Clausen that things were not very encouraging. Clausen was sending an entomologist to Mexico to act as Smith's assistant there, and planned to himself travel to Los Angeles to meet Herb on his arrival there.

Smith was discouraged that he was not having better luck rearing out parasites. He was wondering if the wrong spray gun had been used – one containing DDT instead of another one that had only water to freshen up the leaves. The flies themselves seemed to be okay so Smith thought perhaps there were other reasons. Still he was concerned as the flies were less numerous than expected. Smith tried placing some plants directly in the field under trees but he was concerned they might pick up some disease.

Smith wrote "I am beginning to wish I hadn't taken this assignment. Too many things can happen to ruin the whole work. If only my luck would break."

On top of these problems Herb came down with some sort of flu. The local doctor thought it was dengue fever. Smith tried to remain positive, writing "anyway I'm glad it's not malaria or scrub typhus." A few days later Smith wrote home again.

> *My dengue fever didn't last long, only three days. You feel awfully depressed and miserable. It is better that than malaria or the scrub typhus. The scrub typhus unit is here again now so they can cure*

it in one day. They were away since last winter and none of the drug was available here. I feel a lot better today and worked the whole day without feeling it.

I finally wrote Clausen about the poor showing of the work. I have only seen two parasites so far from all the work I've put in.

Smith was referring to new parasites being reared out in his cages. He was quite discouraged. He was receiving no input at all from Washington and he was getting ready to leave.

Clausen got a number of parasites to come through in his work. He's always lucky that way. I probably spend too much care on them. He didn't even open the cages but once, and I open them all day long to check on them. I don't think I sprayed DDT by mistake. But it's possible as the other sprayer is identical to mine and before I thought of it the boy might have accidentally used it a few times. Yet the one they use is always cleaned out afterwards. This would have killed the adult flies I have but they are not affected yet.

It just goes to show how one must pay attention to the smallest details. Clausen wouldn't have bothered spraying the leaves, I guess I shouldn't have. If I only get a hundred of each species through to Mexico I'll be happy, but to get none would take a lot of explaining.

I still don't know how I'm going to go. The strike is still on and there are no ships to the West Coast now. Clausen thinks I could send an air shipment OK but wants to wait till the man gets down there and set up. That may be about the middle of December. Even the shipping of leaves is a problem as there are not any at the right stage now. I would have to wait till January. Then the infestation is very low here and there are not many trees with it on. Usually I find only a few infested leaves on a tree and lots of trees have nothing. It means hunting around in backyards etc.

Gosh, I'll be glad when this job is done. I have had bad luck since I started. I couldn't find the predator in Cuba after two attempts and now this.

Smith considered the possibility of shipping back collected parasites by air express as an alternative to the massive old-fashioned Wardian cages that Smith was dragging around with him. The parasites emerge over a period of 15 days after the infested leaves are collected. Smith figured that a collection of blackfly from non-cankerous trees could be made and shipped by air via San Francisco to Mexico in about seven days. The cold temperature on the plane might further retard the emergence, which would be good. It seemed promising.

In December Smith was still working long hours, seven days a week until six or seven every night. But things still weren't going all that well. He was washing the leaves in case DDT had accidentally been sprayed on them, but he felt if that had happened it was too late. DDT is a very persistent substance. "Of all things this had to happen - the worst!" he wrote. "It shows what you have to watch. The pest is almost cleaned up here with my collecting all the leaves. It just makes the work that much harder." Herb wrote a follow-up letter on December 16th and he was more discouraged than ever.

> *Well things are no better. I'll just have to struggle through it. I am pretty sure the wrong spray gun was used, at least once. The boy probably got the lab one by mistake, put water in it and it had some DDT in it, enough to kill both my flies and the parasites. The DDT sticks to the leaves for a long time. Three months or more. I have been watering the leaves a lot, trying to get it off, but you can't see it. The parasites walk on it and die soon after. Yet I got a few flies out and they walked on the window all right. So did a few parasites of those I put in cages.*
>
> *I hate to write Clausen that DDT was sprayed. He knows that would be the end of the parasite project. I can't get any flies here, or I would get plants and start over again. I have to go way back to Mexico to get new plants and the fly.*
>
> *To think of all the money I've spent and one little thing – one mistake – ruined it all. I should have thought when I bought the spray, to mark it well and caution the boy about it, also keep it in a special place. I was so busy at the time that I overlooked the danger of using theirs by mistake. Yet I knew they had one and were using it. Probably didn't need the spray at all. But I thought it would freshen up the leaves more.*

I had everything going so well. I collected lots of parasites in the field. Everything went well until this. I would have had a lot of parasites now and been on my way home. Now I'll have to worry all the way back. I have to work much harder to correct the mistake – if it's possible!

I have some material I placed in the field and I have parasites in there, but whether they will breed on the trip I don't know. If they come out they will probably hit the DDT on the leaves, if there is still some there, and I won't get a thing. I will probably write Clausen and let him know what I think the reason for the lack of parasites was.

Everything hinges on this too...A man waiting for me in Mexico etc. Getting parasites over isn't the main job either. It is getting them established there. Strange how one little mistake can affect your whole life. I might have to come right back here and then go on to India. It will be a worry all the time in case something goes wrong again. I may not even have a chance to go home. Or Clausen may get fed up with me and take me off the job. Then I would probably end up back at the New York Port. Well, I can retire in a year or two. Maybe a chicken farm wouldn't be so bad.

In Singapore on New Year's Eve, Smith noted how quiet things were. There didn't seem to be much celebrating going on. He decided against going into the city. He wished he was back in Kuala Lumpur, which was more to his liking. But he had come to Singapore to check on ships headed back stateside. Smith also thought he would find plenty of insects that he was interested in, but he found they were scarce.

> *Drinks are very expensive here. I haven't been to a night club or café yet. I usually go to bed right after supper as I am four miles out of the gay or business section. I haven't seen anything here in the day-time that looks like a night life quarter.*

Herb was also disappointed that the facilities at the Botanical Gardens there were much inferior to those at Kuala Lumpur which had a well-equipped lab and plenty of assistants. The Singapore Botanical Gardens lacked electricity and an entomology library, and the scientists there were all botanists. Furthermore it wasn't easy moving around as he had to drag his heavy Wardian cages and to tend to his insects.

> *Anyway I took a chance thinking I'd find some bugs here and make an air shipment. Now I practically have to loaf for one month and I don't like it. If I had plenty of parasites coming along and knew I would get them to Mexico I wouldn't mind. But with this worry of getting to Mexico with no parasites, loafing here is worse than working, and it costs too much to lug all my cages back to Kuala Lumpur.*

In spite of the difficulties Smith was considering a week-long return trip to Malaysia to try again to find more parasites. He wrote to Clausen indicating that the wrong spray gun may have been used. Smith wrote home that he "knew this was going to be a tough foreign job but I never expected a spray gun to ruin it. And with the money I've had to spend on it, I hate to write up my expense accounts."

Smith wrote home again on January 7th.

> *I have been awfully discouraged at times, spending so much money and maybe get nothing for it. Clausen got some parasites back.*
>
> *Then here at Singapore I have practically nothing to do and that is terrible. I should have stayed at Kuala Lumpur. It was much cheaper there to live too. I may be going back there for a week on January 10th. It depends on if I get the assistant's address in Mexico in time. There are none of my bugs around here. Also the facilities are not as good here as in Kuala Lumpur.*
>
> *I hate to loaf here, but how was I to know how it would be here? I wouldn't mind if I had success with my bugs. Then I would enjoy it here. Just because of a little mistake – spraying with the wrong gun! I always let my assistant spray the cages with water. It is no special job. It was at a time when I was very busy trying to find the parasites in the field and learn something about them.*
>
> *I have had so many worries ever since I took this job but this is the worst. If Clausen doesn't take it too bad I won't mind. It's hard enough to just keep the plants alive and also the pest in the cages. I don't see*

how Clausen did it. He didn't spend so much time on the trip home. Also Havana is easier to get to. I can't go via New York because it is too cold.

The weather is more uncomfortable here than at Kuala Lumpur, damp and hot. Everything costs so much here. I pay $6.25 here for room and board. At Kuala Lumpur I paid $3.50 and had better service and a better crowd. There were a lot of young chaps there and we all ate at the same table. That always cheered me up. Here there is no one to talk to. I never mind being by myself usually if things have gone well. I was alone in Cuba and Panama and enjoyed it.

The English must make a lot of money here if they can afford this place. They are all married and most have babies with nurses. There is nothing to do in Singapore. No gay life or nice restaurants except very expensive ones for foreigners.

A week before his sail date, Smith got word that his assistant Mason had arrived at Colima so he returned to Kuala Lumpur from Singapore to make an air express shipment to Mexico. He wasn't able to make the shipment from Singapore as there were insufficient parasites there. Fortunately the citrus blackfly at Kuala Lumpur was in the proper stage. Leaves with about 300 clusters were collected. Trees around Kuala Lumpur were thoroughly examined but this was all that could be found. The leaves were packed in small cloth sacks with a few tembusu (*Fagraea fragrans*) leaves to conserve moisture.

Manila, 1949. Smith spent many days at Manila while the ship was loading cargo. The devastation from the war that had ended four years earlier was still evident and widespread.

On January 15th Smith returned to Singapore and sailed two days later with the nine Wardian cages on the S.S. Dartmouth Victory. The ship was the first of any line to leave for the United States after a strike. The planned route had stops at Port Swettenham, Belawan in Sumatra, and Penang. Then there were additional stops for cargo in the Philippines at Manila, Cebu, Isabela and Nasipit. The original schedule called for a route through Medan, Hong Kong and Manila abut this was changed while en route. The trip took much longer than expected as a result.

Smith was hoping to do some scouting during the Sumatra stopover, but the ship only arrived at 5:00 p.m.

and left later that night. He checked out the citrus trees around Penang but no infestation was found among the small numbers of citrus trees grown there. Smith wrote home again upon his arrival in Manila on the 29th.

We are now coming into Manila Bay. We have been at sea since Monday night and it is now Saturday morning, which is only five days. It was a little rough the past three days in the China Sea, especially in the stern where my cabin is. You get the up and down movement which sometimes is a lot. I felt sort of groggy for a few days but didn't get sick. It was the nearest I've come to being sick on this whole job.

We don't know what the schedule is after Manila. The ship went aground (before I came on board) so they are taking all the tin out of her at Manila. The insurance company doesn't want to chance such valuable cargo in a damaged ship. We may not go to Hong Kong. Anyway we should know in a day or so. We may load coconut.

I have kept fairly busy cleaning my plants. At night I read a bit. The officers don't sit around much, just go to their rooms. We may get a couple of passengers at Manila so that will be company. If I don't go to Hong Kong there will be little in the way of souvenirs to buy.

The cabin isn't too bad. I have a lounge next to my room with a refrigerator. I put hard boiled eggs inside for night food. We eat dinner at 5:00 P.M. so I get hungry around 11. Bought a few bottles of beer in Penang and keep them in the refrigerator.

They say cigars are excellent in Manila. There is no duty on anything from the Philippines so I can bring in anything duty-free. If we stay at Manila several days I'll look up the Department of Agriculture. I have the name of an Entomologist there and he could show me around the citrus plantations. Maybe I could find some of my parasites. At Penang I looked around but found very little citrus and no infestation.

I expect it will be very hot here in Manila. They all say it is. The weather was comfortable in the China Sea. The room is not so hot and I can sit inside during the day. They have been chipping paint the past three days outside my cabin so I can't sleep in the day time.

We are just coming into Manila Bay and are passing Corregidor where so many Americans lost their lives. It is a small island covered with jungle. You can still see the barracks all burned out and the hospital etc. There are lots of ships still sunk in the harbor, 50 or more sticking up all around. We are anchoring a half mile from shore so will have to go by launch. We'll be here till February 4, no one knows if we go to California then. It's Saturday night, and I thought I'd go ashore for a few beers and to read some US magazines. The launch leaves the ship at 6:30 and returns at midnight.

The ship stayed tied up at Manila day after day. Smith went ashore everyday. Usually he was back on board by

6 PM as there wasn't much of interest in the evening. Saturday night he did accompany some of the men from the ship.

The boys just go up to some beer joint and drink all evening, moving from one to the other. Whiskey is cheap here, about the only thing that is cheap here. Beer is 50¢ a 12 oz. bottle in the bars. In small eating places it is 25¢ but they put a plate, knife and fork down too as you can't get beer without eating. That's the way they get around it.

Smith went out to the Agricultural Department to see the entomologists there and was given a tour of the station. All the original buildings had been destroyed. The next day the entomologists spent the whole day touring nearby orange groves.

We don't know now how long we'll be here, another week probably. I can't get any ship to get there quicker. At least it is cooler here than at Penang. It is boring laying here at anchor. It is hard to get a boat to go ashore. The main launch is at 6 PM and returns at 12 which doesn't interest me. I've gone ashore in the day time but had to wait a longtime for the boat. We are loading scrap iron night and day and it is very slow. All is handled by hand.

The city was all shot up. Practically all the buildings they have are new since the war. It was probably as bad or worse than Berlin. There's nothing to see ashore. They had a lot of nice hotels here before. The Manila Hotel, the biggest, is nice but only half rebuilt. The rest still stands as a skeleton, a depressing sight. There are thousands of jeeps here and they have made them into small buses. They are everywhere! I haven't used any yet because I never know where they go. People don't seem to use them. I have had to use taxis.

I'll be glad once we are underway to California. I like the motion of the ship showing me we're getting somewhere.

It is depressing here, going ashore and seeing the buildings gutted by bombs or shell fire, the piers the same way. Bodies are still in the Japanese war ships, there's one just across from our ship. When the Americans retook the place they caused much of the damage.

The traffic is terrible in the city, with all this jeep business. They sold them the jeeps for $150. Imagine that.

The sailors on the ship get $2.50 a day while they are here because of it being mine infested waters. I'll be on this ship for 40 some days. The man who cleans my room must be ashore because no one has made my bed yet. They stay ashore all night at those "houses." They all make good money – even the mess boys – so they spend it. I'd like to be on another type of ship, some other nationality, to see what they do in their spare time. Here it is drink and women. One chap fell overboard he was so drunk, a Scot, Angus MacLean. I still find more and more of the officers who are divorced. Some the officers are way overdrawn on their *pay – too big an allotment to their wives – so they borrow and don't pay back. I've been warned of the ones who do that.*

One morning Herb went out to the Department of Agriculture and met the men there. Most of their scientific books were destroyed in the war. He went with the director Dr. Merino to a Chinese restaurant for lunch.

His car was broken down or he would have taken me to his home. He only gets about $250 a month but like most officials over here they have outside businesses. He has a couple of big farms and is director of two corporations.

The Filipino people are very nice. They seem to like Americans. They seem to be smart people. From here we go to Cebu for coconut. We will be at the dock there. I hate this being at anchor as you can't go and return when you want.

The next day Dr. Merino drove Herb 130 miles out into the countryside past countless rice fields to visit the Citrus Experiment Station. The roads were good, mostly concrete, with lots of bus traffic. He was disappointed to find no blackfly at the citrus groves there, which meant no parasites. They told Herb the species he was looking for didn't occur here. Smith had found another species of blackfly in Manila, but there were no parasites on it.

We had dinner at the Director's home. They had rice, fried eggs, small fish, and tomato salad. They eat a lot of rice here. For dessert we had sour-sap, a tropical fruit. Then afterwards a cigar.

We went to a chapel where people go to see a new Saint. A nun had a vision about four months ago and afterward roses dropped down from the skies. One of those miracles that often happen in these countries. The doctor with me said he saw the roses himself.

The city was entirely stripped by the Japs when they left. They killed 17,000 people just out of vengeance because they were losing. This is what happened in practically every village and town. The Japs killed a lot of people just because they happened to see them on the street.

We drove through the largest coconut forest in the world. It extends for 60 miles. People make an easy living from their coconuts. The natives seem better off here than in Cuba and more content.

Then we went to the Agricultural College. It was an internment camp for U.S. civilians during the war. Most of the buildings and homes here were also destroyed when the Japs fled. The civilians were liberated by a small group of soldiers and taken across a big lake to Manila. This fooled the Japs as they expected them to go by road where they were waiting for them with machine guns.

There was a rumor that the ship would finally be leaving on the 8th, a week and a half after calling at Manila. Smith heard they would head next to several ports aournd the Philippine islands to pick up loads of copra (dried coconut meat) and lumber. That would mean the ship's arrival in Los Angeles would be delayed until the first part of March.

Smith couldn't wait to get going again. It was boring sitting around with little to do. At anchor it was difficult to get ashore, and once ashore there wasn't much to do except walk around, go to the movies and drink a few beers. On March 7th he wrote home again.

> *We are leaving tonight for Cebu, another port in the Philippines. We will load coconut there. We will probably be there three days, then to another port for a few days. Then another port for lumber. After that we head for California. I didn't get any company, no passengers at Manila. They almost had someone but he couldn't scare up the necessary money to sail so I am still alone and probably will be the whole trip. I bought a lot of magazines but have read most of them already.*
>
> *I was glad to leave Manila. We were there ten days. Being at anchor all the time was very inconvenient. Usually I need to pay $1.00 to get ashore as the company only has one boat at night that is free. Then you have to stay ashore so long waiting for the return launch at 6:30 P.M. This means I spend more money on beer, about the only thing you can do around here except walk or see a movie. I went to the movies and*

saw "The Roadhouse" with Cornell Wilde. They've got a lot of good movie houses here.

The ship arrived at the island of Cebu at 5 AM on the 9th. Cebu was the main coconut port and the freighter was due to load those there. Smith went ashore in the morning and stayed all day.

This city was also badly destroyed by the war. The streets are dirty and dusty. They have a few good movie houses. The stores are practically all Chinese. I went to a movie in the afternoon, saw "The Message to Garcia" with Wallace Beary.

From here we go to two more ports and then return here. It may be March 5th or 6th before I get to California. I will have had enough of ships to last a lifetime.

Herb looked up the local Quarantine Officer who took him out to a nearby nursery to inspect citrus trees. They were free of blackfly. In the town, Smith did find one lime tree heavily infested with the citrus blackfly, but he saw no parasites. He was still hoping to locate more parasites to introduce into the massive Wardian cages he had been lugging all around Southeast Asia. The caged blackflies were at the right stage of life for parasitization, but Smith was having no luck.

The ship left Cebu a few days later, sailing for the port of Isabela on Basilan Island. Isabela is one of the poorest cities in the Philippines and is now a center of activity for

Abu Sayyaf Muslim extremists. Smith, however, found it on his visit to be a sleepy port "with only three little bars." He walked around the town, taking in a few movies. "The islands here are covered in coconuts," he wrote. "It's hard to find something to do here, but with this heat you don't feel like doing anything."

After three days of loading coconuts, the ship again raised anchor and returned to Cebu. Smith noted "I've already been on this ship for one month now and am not far from where I started - Singapore." He kept checking for any faster options for getting from Manila back to the States but there was nothing. The ship stayed out at anchor, meaning each trip to and from shore required waiting for the official launch, which only ran at 6:30 in the morning and at midnight.

The local government officials continued to be helpful, taking him around the area.

> *I went ashore yesterday morning and saw the agricultural chap. He took me out for a couple of hours in the afternoon. I still didn't find any of my bugs, just a few leaves, and of course no parasites. Anyway I saw a good bit of the country for nothing. I came back to the ship at 6 PM. I spend most of the time on shore walking around looking in shop windows or sitting in different beer parlors reading magazines or papers.*
>
> *Today I had an appointment again with the agricultural fellow at 8:30 AM, but he wasn't able to get the station wagon until 1:00 PM. We went out to another part of the island and looked at more citrus*

trees. I didn't find any bugs this time, either. So I have given it up.

It wasn't certain how long they would be laid up in Cebu this time. But the next day they heard they would be sailing for Nasipit that evening so Smith went ashore "for a last fling."

I was drinking beer in a place with the first cook, a colored fellow. The stewards are all colored. The men all mix with them and drink with them ashore. Officers and everybody mix because they are all from the West Coast. This fellow has been around a lot more than I have.

In these bars in the Philippines you order beer and a girl brings it to you and sits down with you for company. They don't drink anything. You just talk to them. They get 10 cents for every bottle you drink but they don't rush you. They have a radio or juke box and dance their native dances. Some of the men off the ship marry these girls. They all speak English fairly well.

Sometimes I dance when I've had enough beers. They still do the jitter-bug. There are about ten bars like this on the waterfront and you always find the men off the ship there. The girls are always pleasant and listen to your tales.

The common form of transportation was five cent rides "in little cars run by ponies," as the taxis are too expensive. Herb met a local laundry man, an American who had been in Cebu for 20 years. This man was a major with the guerrillas during the war and had made money in gold mining. He told Smith that he had twenty children and was thinking of marrying his "wife."

The next morning the ship made port at Nasipit to pick up a load of lumber before heading out into the Pacific towards Los Angeles. The ship arrived at Los Angeles March 8th, completing a voyage of 50 days from Singapore.

At Los Angeles Mr. Clemens, in charge of the California Plant Quarantine Station at San Pedro, met the boat and arranged for storage of the cages at his headquarters while waiting transfer to a ship going to Manzanillo, Mexico. Smith found returning to America after many months abroad to be a bit of a shock.

> *Things seem so different here. It's still like a foreign country as I am not at home. But here I have a good old USA hotel room, carpet on the floor, plenty of furniture, cupboards, lights, heat, double bed, and shower – almost like home. So different from where I've been. I can't adjust myself even though I've been through it dozens of times. It always seems like a new experience getting back to the US.*
>
> *I felt blue all day over my bugs, even though Clausen never said anything, and seemed to be in good spirits, as though everything was fine. To him it probably was.*

My air shipment came through in good shape even though delayed in Mexico, the fault of my man there (two days). Clausen is an optimist about things which is probably why he is where he is. I look on the dark side too often.

Anyway the quarantine men really did themselves proud over me, like I was visiting royalty. First the boarding officer visited with me at 7:30 AM. He knew about my coming. Then when we docked another inspector met me with a car and drove me to the office.

In the meantime the chief and Clausen and two or three other men came to the ship looking for me by another route. So when the other chap and I landed at the office everybody else was at the ship looking for me. We just waited and he showed me around, and finally the rest of them showed up. A real swell affair! I wish I could have measured up to the importance. I wish I didn't get so nervous and worry so much about the job.

The next day Smith went back to the ship to arrange for his Wardian cages to be brought to the dock. A truck would arrive the following day to carry them to the Los Angeles lab for safekeeping until Herb could arrange passage to Mexico. Clausen came around for a talk. Smith was still fretting about the incident with the spray gun, but Clausen never raised the matter.

Smith had frequent dealings with the entomologists in the Department of Biological Control at the University of California. The group is pictured here at a 1948 gathering. The university operated the state's Citrus Experimental Station in Riverside. Smith spent several days there after returning from Malaya in 1949, consulting with Stanley Flanders (lower far left), Wendell Sellers (man in center with black trousers), and Harold Compere (balding man to the right of Sellers). (Photo courtesy of the University of California.)

While in Los Angeles Smith spent most of his time meeting with the entomologists out at the Citrus Experiment Station at Riverside. There was Stanley Flanders, Harold Compere, and his old colleague from Paris Wendell Sellers among others. Herb found the talks on parasite technique and biology to be very useful. A trip was made with Dr. Flanders to the predator and parasite breeding stations around Santa Paula.

On March 30th the cages were placed on board the Swedish ship S.S. Glimmaren, the first available ship, As no passenger accommodations could be secured on the ship, Smith left by air the next evening for Guadalajara and thence by train to Manzanillo.

8

Mexico: Big Country

Smith flew to Guadalajara on a small plane with five other Mexican passengers, a pilot and a steward. The five hour night flight traveled over rugged mountains but he felt "at night when I couldn't see anything I didn't worry about things like what would happen if one motor conked out."

After breezing through customs at the little airport, Smith waited a while for his luggage before thinking to show his tags to the clerk. Back in Los Angeles they had marked his bags "Mexico," which unfortunately meant they were still on the plane just now at the far end of the field readying for takeoff to Mexico City. A quick call held the plane and a station wagon was quickly sent over the retrieve the bags. Smith wrote home again.

> *I'd have hated to see them go to Mexico City as I may never have seen them again – all my checks were inside along with my typewriter. Funny what close calls one has and I usually check and double-check so that people think I'm a nuisance. I don't take one person's answer as final. Maybe that's why I've had such little trouble so far. I should have watched for my things coming off the plane and I did for a while but there was a big gathering there with a band and so on to welcome some Mexican big-shot so I watched that instead.*

After checking into his hotel, the local Vice Consul invited Herb to his home that evening. There were drinks but no food, as it was their custom to eat dinner in the afternoon. Herb found somewhere afterwards to get some fried eggs and beer. The next day Smith went to see the local Mexican entomologist.

> *He is working on the blackfly. It is very bad here. Today he took me out to some farms to see the citrus trees. Gosh, I wish my blackfly had grown as well as it does here, leaves are black with them, and no rain for months. I watered my leaves all the time. Maybe I shouldn't have bothered, but according to Clausen's notes they need water. He is wrong again. The parasite doesn't do well here though. Tonight I leave for Manzanillo. I have a sleeper. I didn't want the day train. It is too slow and dusty.*

Herb Smith at work at his microscope in Mexico, 1950s (Smith family collection).

Smith arrived in the Port of Manzanillo the morning of April 3rd. He waited there for his Wardian cages to arrive on the Glimmaren. The ship arrived the next day and the cages, which had been stowed on the afterdeck in a wooden enclosure, were in excellent condition.

Two days later Smith flew with the nine cages to Colima. He found that rooms were much harder to find there than at the more touristy Manzanillo. "This is about the worst hotel I've been in with the poorest food," Smith wrote. "No decent restaurants in town. We don't have much for a lab either."

Headquarters had sent an entomologist called Mason to Mexico to assist with the blackfly work, Mason and his wife traveled two hundred miles over the mountains to meet up with Herb in Colima. It took them 22 hours over poor roads that often disappeared into dirt tracks where one had to back up half a mile to let trucks pass.

The cages were kept in a room there for the next two months and checked for the emergence of any parasites. Some Eretmocerus were recovered but no *Prospaltella divergens* or *P. smithi* emerged, The plants were burned and the cages cleaned, shellacked and stored at the Hacienda Albarradita until needed again. Smith was already beginning to plan for a follow-up trip to India to search for parasites.

Smith was discouraged and doubtful about the prospects in Mexico, writing:

> *You just can't believe how primitive it is here. Gosh, in Malaya I could ride anywhere, even in the*

Philippines. I can't get out of here too soon, unless of course I get some parasites out, that would cheer me up and be different. I doubt if I'll get any out. Mason didn't handle them too well, but it's not his fault. He only had a 10-power hand lens and he knew nothing about parasites. He never went to Washington and never saw Clausen. Clausen never told him anything. But he did the best he could. It's going to be hard to send the air shipments here as the Mexican authorities hold them up at customs. Things just don't go off smoothly here.

Meanwhile back in Washington, the Secretary of Agriculture Charles Brannan asked Congress to appropriate another $150,000 a year to assist Mexico in dealing with the citrus blackfly. He noted in one Mexican area of 4,000 citrus trees yielded only 15 boxes of fruit last year, while the same area produced 7,500 boxes in the 1940-41 season. He told the Congressmen the infestations were within 250 miles of important citrus growing areas of Arizona and California.

Herb next decided to make another trip to Cuba to see if he might have any luck finding the elusive predator *Catana clauseni*. He left for Havana on the 26th of May. Smith hired Señor Otero, who was involved in the original releases of Catana in Cuba in the 1930s. The entomologist F. Valdes Barry from the Experimental Station at Santiago de las Vegas accompanied the pair from time to time. Transportation was via a jeep borrowed from the Station, by rented car and by taxi.

In Havana Herb received instructions that he would need to go to India by boat. He had wanted to travel there by airline, without the 300 pound Wardian cages this time, and to ship back parasites and predators that he collected via air freight. Clausen agreed that this was the best plan.

There was a conference about it back at headquarters and the Plant Pathology department objected. They were concerned citrus leaves airmailed from South Asia to Mexico could have undetectable citrus canker. If that were to be inadvertently introduced into Mexico, that would be a big problem and perhaps an international incident.

This was a disappointment in two respects. First, the best results from Smith's Malaya work were with air shipment rather than the slow ocean voyage with the cages. The second was that with the Wardian cages to attend to on the trip over, he would be unable to make a side trip that he had planned to Paris to visit with his daughter Janet and son-in-law Jean-Jacques and their children.

The blackfly in Cuba was much less abundant on citrus than it had been when he was searching there the year before. As a result Smith focused on papaya and bamboo. The papaya whitefly was widespread near spot where Catana had earlier been released, but the whitefly parasite Catana was still nowhere to be found.

Returning to Mexico in mid-June, he wrote home describing his activities.

I flew here this morning in a little plane. It had two motors though. I was the only passenger, only room for six. The pilot was just a kid – about 21. He motioned me to sit in the co-pilot's seat from Colima to here.

I got back from Cuba last Sunday. I stayed at Mexico City three days. Everyone invited me out till I was so filled up I was glad to get away. They are a nice crowd there. I enjoyed myself there at the lab. Back with men who are like myself. They are honest, and straightforward, and work hard, and put in about 9 hour days. They even let me read their notes and reports etc.

Not like this man Mason. I told them about Mason and they suspected he was that type. They think I got a raw deal with him. It's his fault I had to fly here today for more plants. He only looked at my citrus once while I was away. Many died. They told me to give him the lowest efficiency rating, and I probably will. This is poor. It means that he has to be fired. He didn't infest my trees at all.

I may have to spend four more months here because of his laziness and stupidity. Then Clausen may do something about him. I told Clausen he was worthless but he said he couldn't get anyone else. I can get someone! The chap I had in Cuba was 100% better and no college man. The 16 year old boy is my assistant; he knows more about the work than Mason.

I just let Mason stay in his house. It makes me mad to think that good American money is paying him to loaf. The boy said he only looked at my plants once while I was away for more than two weeks. He only went to the citrus groves once. He should have told me the trees were dying in Mexico City, then I could have picked more trees up on my way. Now I am delayed in getting to Washington and home. I'll tell Clausen about Mason and put it to him strong. I can find men better than he is. He goes around thinking he is a big shot but won't turn his hand to a thing.

It was a pleasure to be in Cuba with men who work and have ability. Then I come back and run into this numskull. I don't have anything to do with him as I can't make use of him. I had him drive the car today, while I'm here, to pick up some leaves collected by boys. That's the first work he's done in weeks.

I'll be back in Colima Friday. Then I'll have to get my stuff ready for Washington. I have lots of specimens for the Museum.

I didn't find that bug in Cuba. I guess it has died out. I wish I could get some successes once in a while. Even my plants are dying and they never did in Cuba last trip. Well, I am getting more tomorrow so I hope they will be OK. Tomorrow I'll have nothing to do here. I hate to loaf. I'm happy when I'm busy. If I could only have the conscience of Mason – he does nothing but read and write letters, and doesn't worry about the money he is paid for doing nothing.

Smith left Colima for Washington on the 19th. After a few days of consultations there, Herb returned home on his annual leave. After a month of rest and recuperation, Smith returned to Washington for ten days to prepare for his trip to India. He then flew to Mexico City on August 8th, then after a day continued on to Colima.

From the 19th to 21st of August Smith was in Tampico awaiting arrival of the cages and securing more seedling trees. He was also going around inspecting the control work at Valles and attending to various matters necessary for embarkation.

Smith left Tampico August 21 on the S. S. Agwi Princess and arrived at Baltimore August 27. There he disembarked there but left the cages on board to be discharged at New York. The next three weeks were spent at New York and Hoboken waiting for the next available ship to Pakistan. Finally on September 18 Herb sailed on the S. S. Exchange, with an expected landfall at Karachi around October 10th.

While at sea, emergence of the flies in the cages continued with a peak on around September 22nd. A heavy infestation was secured on the Tampico seedlings and most of the mangoes. The Guadalajara citrus had no new growth at this time but the infestation on the old leaves was satisfactory. At the end of September many new shoots began to appear on these plants, the first since potting them four months ago. Adult blackflies were present in moderate numbers during all of September thus assuring a good overlapping of generations.

If Herb was experiencing a sense of deja vu, that would have been understandable. For the second time in two years, Smith was shepherding a ton and a half on large wooden Wardian cages with their cargo of citrus trees and citrus blackflies across the Atlantic and the Mediterranean, through the Suez Canal, and into the Indian Ocean. He was hopeful things would go better this time.

9

India and Pakistan: On the Insect Trail

Herbert Smith arrived at Karachi, Pakistan, on October 10, 1949. It took several days at this port to arrange for the storage of the nine Wardian cages of citrus plants and to meet with the various local officials. Smith then left for Lahore, and upon arriving there he spent a week surveying the area. Infestations by the blackfly *A. woglumi* were very light with most groves clean. A few heavily shaded trees in one grove were found to have a moderate infestation. *Amitus* was the only parasite found with a parasitization of about 70 percent. Smith wrote home about his first impressions of the region.

I arrived here last night at 9 PM after traveling 26 hours. This is 766 miles inland and near the Himalayas. All the way was as flat as a pancake. Lots of places it looked like a desert. It rarely rains here

Herbert Smith traveled thousands of miles around India and Pakistan over a seven-month period in 1949-1950, often by train and ox cart. He studied the citrus blackfly and its native parasites and predator insects that tended to keep the pest under contol on the subcontinent.

so they can only grow things by irrigation from the Indus River. It was quite hot but I had an air conditioned compartment for four. The others were all Pakistanis, big government officials. They of course all speak perfect English. They were very nice. More like American than British, more sociable. One, at least, went to Oxford. I don't know about the others.

In India you are supposed to always travel with a bed roll (blankets, mattress, sheets, pillow) but I didn't know about it and no one at the Embassy said anything, yet they advised me to go "air-conditioned" as they call it. As it was, I arrived at the train and here all the other men had their bed rolls! I wondered what was up. I thought they were going camping somewhere. They were surprised I didn't know about it.

One chap loaned me a US Army blanket, sheet and pillow, so I was OK. There is no corridor on the train (as in England) so you have to wait till the train stops before you can move. However there is a toilet and face bowl attached to each compartment. To go to the restaurant car the train had to stop, but most of the men had their meals right in the compartment as the restaurant car was not air-conditioned,

The main comfort is freedom from dust. There is an awful lot of dust blown up – everything being dry and burned up and it gets in everywhere – except in an air conditioned coach. There is only one air conditioned coach.

I could have rented a bed roll from Cooke for $3.00 if I had known in time. The conductor said the

(Right) Smith visiting the Government Botanical Gardens at Saharanpur, India, December 10, 1949 (Smith family collection).

(Bottom) Herb took a rare break from field work for a quick visit to the Taj Mahal (Smith family collection).

water would be bad and advised me to buy a native water jug for 15¢ and fill it with water at Karachi station. I did this but never drank any. Most people carry their own water too, as you can't drink the soft drinks here. No labels on the bottles and it might be anything. So I am back to guzzling tea.

The engine ran out of oil before we got here. The driver forgot to measure it at the last station, and you can't get oil everywhere. They burn oil instead of coal as they have no coal in the country. The Chief Engineer of the railroad happened to be in our compartment and was he embarrassed. They had to send back to Lahore for another engine and it had to back all the way at 25 miles an hour. They had no turntable or Y at the station. This country is young and proud and they are very touchy about these things. We waited and waited. The other two Pakistanis rubbed the life out of the Engineer. They talked English mostly but would often break it up with their own language which is – Ordo – an off-shoot of Arabic.

The Moslems of course are not supposed to drink because of their religion. One man did and we had two or three beers in the restaurant car. He was the one who loaned me the bedding. He learned to drink while in the British Army. He came out of the Army to find his mother and father had been murdered by the Hindus – you know, during the massacre all over India at the time of the separation. His three brothers are still living. They were rich before and after the separation he had nothing. Now he seems well-fixed.

Beer was $1.00 a quart bottle on the train and yet he bought some for a friend too. At the Hotel Karachi I paid $1.50 for a bottle of beer and a tip. I got beer for 65¢ once in a store. They have no bars here. They only sell it in hotels. Ale and beer is nowhere near as good as ours. It is shipped from England and they put quinine in it to keep it.

After we got into the Punjab Province we could not get beer at all. Prohibition here. I have to get a permit to buy it and must get it in a store. You can get sixty bottles a month of beer or six of whiskey. Moslems can't get it without a doctor's prescription.

The consul here at Lahore met me at the train with his new Chevy and took all my luggage to the hotel he had reserved. A swell-looking place – deep Persian rugs all over, swell lobby and dining room but lousy rooms. I have to use a "chants" or potty for a toilet. No sewers here. No plumbing. I don't exactly like that, especially when it stinks up the place. Terrible! They have water for baths and so forth though. Karachi was much better. Flush toilets, shower and good beds. Here the bed sags in the middle and I use my old clothes rolled up like a pillow.

When I arrived, the room was full of mosquitoes. I couldn't see any of the malarial kind but there are lots of them here and much malaria. I started my Paludrine pills today. Anyway I had a netting rigged up but it was full of holes. I patched most of these up with tape. They must be funny mosquitoes because they weren't attracted to me at all. None were inside the netting this morning and none were even on the outside when I put on the light at night. This morning they were all at the screens trying to get out so I flitted them away. I guess they only like Pakistanis.

I went to a nearby cabaret last night. Guess what I drank. A pot of tea – three cups. That is what everyone was drinking. There was very little dancing, two to three couples only. Most just sat and watched. It was mostly men there. They played "I am Always Blowing Bubbles."

The officials have been very nice to me. They arranged to meet others at various places, etc. They gave me a lot of attention. I am the first American scientist here, I guess. Quite an honor. They are looking after my cages too.

India-Hindustan will be much better than here. Bombay is a big city. There is lots to see and do there. Here there are no historical places. There is nothing to do but work and stay in my room. The cities are dirty and there are no tea parlors, bars or anywhere to sit down. The shops are small. Everything is expensive.

I looked around for the blackfly but found very little and no parasites. I will know more later when I meet the entomologists in this area.

I took a ride through the old city this morning. Not very exciting. Cairo was much better. Just a lot of junky shops. Taxis are expensive. Gasoline is 75¢ a gallon. There are very few bars here. Malaya was a paradise compared to here. This is the lowest I've ever been. Anyway I have a room to myself. I had to share in Karachi. The food is very ordinary. They must make a lot of money on foreigners. When I think of the meals we have at home. One would be a banquet here.

Yet the British (only a few remaining) lap it up and say nothing. The coffee is a half teaspoon of coffee in two cups of water that is allowed to steep.

My cages are handled by camels. They have few trucks here. One big camel took the whole lot. They say they like cool weather, and when it is cool they get frisky and bite people. But when it is hot like it is now they don't. Just like human beings, heat takes the pep out of them.

The air service is not too safe. They use old surplus army (USA) equipment – no service, etc. Even the Pakistani officials don't use them and advised me against it. One chap in my compartment took 18 hours from London to Karachi and 26 hours from Karachi to Lahore by train. If planes were any good he would have taken one. One plane blew up in the air a few months back. Yet it is as cheap to travel by plane. No mountains to bother them, all flat country.

I am eating the usual English meals again. Early tea at 7, and breakfast at 8 or earlier consisting of fruit, porridge, fried eggs (no bacon) and watered coffee. We also had fish. Lunch at 1 to 3 and dinner at 8 to 10. I usually skip the afternoon tea. I drink tea but don't like it. I guess I can drink anything now except poison.

You see very few women here. Those you do see are veiled. This is the biggest Mohammedan country in the world – population etc. I see the veiled women looking through their lace window and wonder what they can make out. It sure is a man's country!

Next Smith traveled to Lyallpur where he spent another two weeks surveying the various nearby regions. This region was a commercial citrus growing area. The citrus blackfly was not found there at all even though it was only 90 miles from Lahore, which did have the blackfly. Only the orange spiny whitefly (*A. husaini*) was found in the Lyallpur region. Some of the citrus trees there were badly infested with this closely related species, sometimes enough to cause economic damage. Smith hypothesized the absence of *woglumi* could be due to the extremely high temperature (120° F.) in May and June.

Have been in this place now since a week ago. This is really out in the sticks. There is no hotel in the city that a European could stay at. I am at the Rest house of the Agriculture College. I pay no rent the first two weeks but I seem to spend a lot of money for meals etc. Everyone (Europeans) has many servants here and takes them about with them. That was the first question the President of the college asked me – if I had a servant with me. Anyway they have a man who cooks my meals, whatever I order. Having no servant I have to make my own bed etc. They clean the room once in a while. I have a small apartment – two big rooms and a wash room, with that ever present potty on a stand.

I bathe the Indian way – take water from the tub and pour it over me. The two rooms are bigger than our living room and have over stuffed furniture, a big table, etc. Not bad. Lots of chairs. The bed has canvas straps, no springs or mattress. The floor has a cotton rug. I am within a stones throw of the orange trees and anything that way is handy.

The city is 1½ miles away and there is no way to get there except by bicycle or horse cart. There is nothing to do there anyway. No bars, of course, and movies once a week. Sunday mornings you can buy beer in bottles with a permit – which I have. Also six bottles of whiskey a month at $8.00.

The grounds around here are fairly nice. There are grass tennis courts across the road and they are always playing there and seem good at it. When I get more time I'll match them.

I am quite a curiosity in town. I stop traffic whenever I take a picture. Many want their pictures taken in the shops. There are no cars here – three or four at most, and for a city of 200,000 people this isn't too much. No real stores – just holes in the wall.

Stuff is laid out on the sidewalks. It takes me back centuries. Lahore was more modern and had a number of English people. Here I guess I'm the only European.

My man makes me lots of curried mutton etc. Now I am getting on better and have gotten water buffalo steak. He speaks a little English. The hotel at Lahore was very expensive, more than my per diem to live there. I never had that happen before. And condi-

tions were poor as well, the food was rather ordinary and the rooms like a fifth rate U.S. hotel.

The Entomologist here has been very nice and has gone around with me. He speaks good English. I have had to get used to drinking water and eating in villages, although I don't do it if I can help it. So far I feel all right.

I found some new parasites at Lahore and also here. I have wired Gardner that I can now ship so when he wires that he has left I'll start getting ready. I will take the first box to Karachi myself by air. The airline isn't so bad – so some say. I will bring back a couple of my cages. If I took the train I would have to rent a bed roll, and the ride is a long and dusty 24 hour trip. It's not a Pullman and it is not too comfortable. Air costs the same amount and is a lot faster.

The President wants me to speak to the college students, 350 of them, but I may duck out of it as he said if I didn't want to, it would be alright. They have a very good college here, as good as ours in many ways. The professors are very good, and their equipment is O.K. They are up on all the modern ways.

Sunday and Monday I go out to a village and see the real life, with the Entomologist. He had me to dinner, quite a treat. The food is very different from ours. You don't meet the wife. They observe "purdah" – wife veiled in the streets. The fact is you see practically no women here. You would hardly know they existed. A man never goes out with his wife, only to visit sometimes. You never see men and women in the streets together.

The children scrape up the horse manure from the streets with their hands and put it in baskets. Then they carry it on their heads to home and make cakes which are dried for fire wood to burn.

They have little wood here and no coal. When one gets a grilled steak here it is not charcoal grilled but horse manure grilled, at least it could be. My man uses charcoal here so I am O.K.

The climate is fine. It is 60° at night and 85 to 90°F. in the daytime. It is always sunny and there is no wind. No one ever talks about the weather – it never changes. The land is flat. Although we are 200 miles from the Himalayas and Mt. Everest, I cannot see them.

I am getting to like it here. My work is handy. I have lots of room and my lab is in my room. Not like Kuala Lumpur where I had to go three miles to work. I have early tea at 6:30 and breakfast at 7:30. That early tea is nice but you have to have a servant as you get it in bed. The flies in the town are so thick that I can't see the food or fruit. But I'm getting used to that now.

From November 7 to 16 Smith was in Karachi and Mirpurkhas. At Karachi citrus was only found in gardens or backyards. Infestations were light and irregular. The only parasite found was *Amitus hesperidum*. Also a species of the coccinellid predator Catana (probably parcestosa), was found in small numbers.

At Mirpurkhas, on the edge of the Great Indian desert, there were many commercial citrus groves, all under irrigation. There was a heavy infestation by *A. woglumi* here, especially on grapefruit. The degree of infestation was similar to that found at Colima, Mexico. The leaves were black with sooty mold. *Amitus hesperidium* was found there, but was apparently offering incomplete control of the pest. Catena adults were also found in fair numbers feeding on *A. woglumi* eggs.

The agricultural officials there reported considerable damage by the blackfly during the past 10 or 15 years. Smith felt their practice of root-shocking the trees may have been harming the various blackfly parasites and the Catena predator more than the blackfly itself. In January and February, most of the leaves fell due to this treatment.

After another two weeks in the Lyallpur and Lahore regions, Smith traveled around northern India for another half a month, stopping to survey in the areas around Agra, Delhi, Saharanpur and Dehra Dun. At Agra, no commercial plantings were found, though a few citrus gardens showed some infestation by *A. husaini* that was parasitized by an *Eretmocerus* species.

The Delhi region had a fair amount of citrus in heavily shaded gardens. The trees looked very poor, and the citrus blackfly had much less than 1% infestation. Only 15 clusters were found in several days hunting. *A. husaini* was fairly common and generally distributed.

Saharanpur was a commercial citrus growing region. Many large groves showed no infestation by either *A. woglumi* or *A. husaini*. A few grapefruit trees in the Bo-

(Top and bottom) The citrus blackly (*Aleurocanthus woglumi*) is native to Asia but is now found in Latin America. If uncontrolled it is one of the most damaging of insect pests of citrus.

(Facing page) Three of the parasites Smith brought back from Asia to control the blackfly in Mexico.

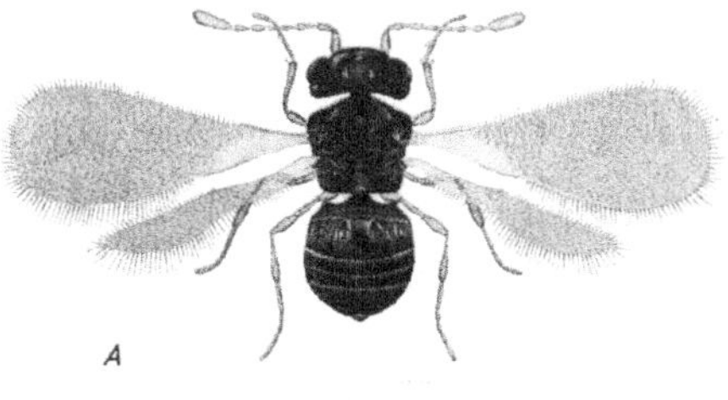

Amitus hesperidum

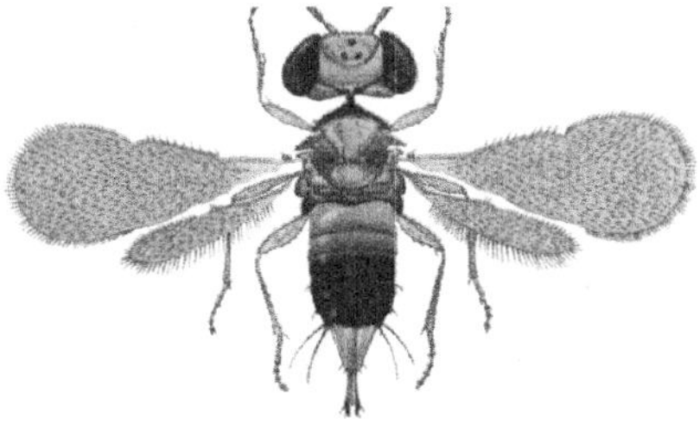

Prospaltella opulenta

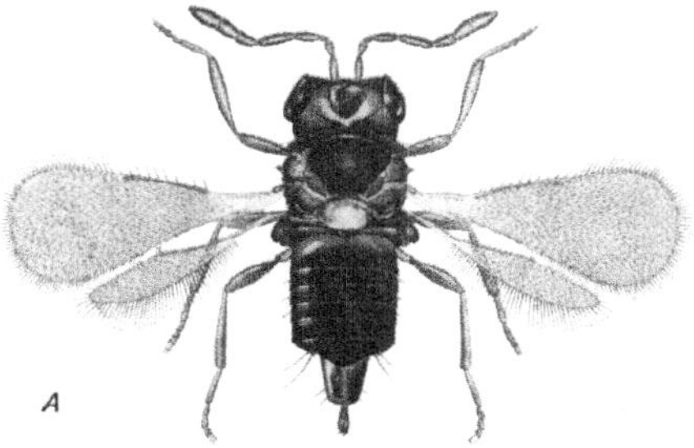

Prospaltella clypealis

tanic Gardens had a fair infestation by the citrus blackfly, though *A. husaini* was much more prevalent. *A. husaini* was found in large numbers in the previous generation on leechi trees. Mango trees had small numbers of *A. husaini* on their leaves, with a 60% parasitization by *Eretmocerus* species.

Smith went back to Lyallpur and Lahore from December 15 to 21. From there he wrote to his family to let them know how the work was going and how he was faring.

> *Gosh, I have been on the go! I wish I could get a break and rest a bit. It is so cold here and the rooms are like barns. There is nothing to do but go to bed after supper. In Delhi I slept in a big ballroom in the hotel, with no place to put my things. I was much better off in Greece. Here at Lahore I have a room – no heat of course. It is 60° in the room so that isn't bad. In Delhi it was 50°. In Kuala Lumpur it was warm and nice – just what I want.*
>
> *I got word from Ted that my first shipment was held in Mexico City for six days before being delivered and most of my material was dead!! Ted hasn't been around as I have and doesn't know you can't trust people in foreign countries. He should have haunted that American Airlines office every day. He just waited in the lab for them to call him up probably. Some clerk kept it on his desk for six days. Ted wired New York and they said it was shipped six days ago.*

Boy! I check and check and check these things. I'll bet Clausen doesn't say a thing to him. Ted knew from my cable that they were shipped November third. I told him it takes three to five days. He should have gone down and looked around the office every day after three days. I had given him notice to line things up – buy the clerks some cigars or a drink and get acquainted.

This would never have happened to me. At Karachi they missed one day on my shipment. A man kept it in his desk. But I was right after it till it left. The American way of thinking doesn't work in foreign countries except maybe England.

Ted must have felt pretty mad to have sat around for six days while the box was right there in the Airlines office. He has a car. He should have gone to the office and pestered them. I won't tell him this of course. He is lots better than that Mason. He has never been in Spanish countries before and doesn't know the ropes. But you would think the other men would have helped him.

You may not receive all my letters as they take stamps off letters here and destroy them. I usually try to get the stamp canceled, but not always.

Ted says he thinks the parasite work on black fly will go on for several years. A Congressional Committee was down there looking into it and a big man in California is behind it. He is head of the $100,000,000 Fruit Growers Exchange. The Chief told Gardner to *shoot-the-works to the Congress men when they came down.*

You see my findings here are all new. The parasites here are many and are controlling the pest. We knew absolutely nothing of the parasites here and even of the pest itself until I came here. Now they will study my reports and act accordingly.

Clausen doesn't write much because it is mostly over his head now – that is – he is not in touch with this type of work now as I am and just writes general letters. He has no suggestions to offer. I am definitely on my own and with no help. I know more about the parasites and the pest and so forth than he does, so he doesn't stick his neck out.

I have been feeling OK, no more colds. I am awfully fed up with village cooking. I shudder at the thought. Greasy mutton in hot pepper day and night. In cities I get better food. I took a bath, the first in three weeks. I haven't washed my hair in over a month.

Boy, was I lucky! The Pak Airline had a crash at Karachi on December 12th (26 people killed). I hope you didn't think I was on it. I saw the notice the first time today when I took the plane from Delhi to here. It was the Indian Line. I was worried a bit about that plane. It was the same type that crashed. I guess my luck still holds. I was a bit worried about the Pak airplanes but took it just the same. Now they are closed up everywhere. The worst accident in Pakistan. They had a day of mourning – all stores closed. Big men lost their lives.

No more plane travel for me here. In India they are better. The pilot when I was on the Pak plane just started the engine and flew away. Usually they warm it up. After that I decided not to use the line. When I think of what could have happened.

I'll be going to Karachi next week and spend Christmas there. What a place! Then to Bombay, India. I gave a talk to 50 people in Delhi. It was well received. Now I have to talk again at Lyallpur.

Smith was extremely busy and working long hours. He hoped to get a break to have a few days to relax and to send in his expense account for November. On December 20th Smith wrote another letter to relatives.

I am still too busy to like it. I almost lost my note book with all the "dope" in it the other day. I went to a chemist and left it on the counter. Of course I didn't know where I left it. I found it was missing when I got back at the Rest House several hours later. It might have been anywhere. I rushed back at night sitting on the cross bar of a man's bicycle. My servant rode the bicycle for miles – no light – and it was very dark and cold.

When I got into town I hailed a turga (horse cart with two wheels) and went to the chemist's. He was closed. I had a job locating him but he had my note book with my December expenses and all my Pakistan notes and addresses. I'd have had an awful job remembering what was in it.

I got in here last night at 10 P.M. My special place was closed. I had a turga and it was loaded down with 9 big pieces of baggage. I finally went to the "rich" man's hotel. Poor food there though.

I sent off my last shipment from here today. It is the 8th shipment. I go to Karachi tomorrow, but stop first at a town in the desert 45 miles off the main road. That night, or rather morning, I will get to Karachi. I may have to sleep in the station on the floor. I love my bed roll.

The court is sitting on the Pakistan Airlines plane crash. I notice that the British check pilot had refused qualification for the pilot that had the crash. He wasn't good enough to pilot a plane. When I flew up here by Pak they had a strike that morning and I was delayed one hour. The Pakistan pilots wanted to take over from the British and Americans. But the British pilot told me they refused as they hadn't had the flying time that we would have. Also they didn't try to help or run the planes. Then this unqualified pilot takes a plane up with all the biggest shots in Pakistan and Egypt. It serves them jolly well right!

After I landed I knew I could not travel on that line again. So I took Indian Airlines to Delhi. The Indian people are much more qualified and superior in every way.

I am glad to be getting to warm climates. It isn't cold here but you feel it because they don't have central heating, It's like Scotland, only worse, as the ceilings are 30 feet high. It costs $1.80/day to heat one *room and it wasn't heated unless you sat by the fire and smoked all the time.*

It gets to 38°F at night and is 70° in the daytime. It is always sunny. You need no top coat in the day-time. I had my first hair wash in over a month today. With a nice house it would be swell here. The days are really warm and nice.

I gave a talk to the Punjab Agricultural College students. We sat on the grass as it was too cold in the rooms. I just talked extemporaneously. They lapped it up. All said it was very interesting – on biological control and quarantine work. I talked for an hour. It was easy to talk to them as they know much less than I do. They took a lot of notes. I had a few professional photos taken.

I apparently am a big shot to them. Anyway I certainly have impressed them as I have questioned them in different ways to sound them out and it all measures up to that. While talking I looked around to see how they were taking it and all their eyes were glued on me. I know when I listen to someone and it gets slow I wander away. They would have had me talk on and on. I told them about my experiences etc.

They saw me off on the train and one chap who was coming to Lahore wanted to see me again. You would think I was a Mahatma Gandhi or something. They shake hands and put both their hands about mine – a sign of high respect here. The scientists do it. It is sort of embarrassing as I'm no Doctor and they know it.

They are all surprised how hard I work and get around. No English ever did that. He never made a move without a lot of assistants. Also he never went into the fields as I do. I was told that also in India. That is why the work published here is no good. The big men send the small men out to study things and publish their reports under their name. It is no wonder Washington doesn't believe anything from here. I went out and studied it in the field as we do in the U.S. That shocked the people here. You cannot send a boy on a man's errand and that is what they have been doing here for centuries. We Americans don't do that. To get scientific information you have to go yourself, as I found out.

I am taking the train tomorrow. I will arrive in Karachi Friday and spend Christmas there. What a place. I would have preferred Bombay. India is better than here so I am getting the worst over first. Oh, to get back to the tropics – Bombay is tropics.

Smith hit the road again, spending another 10 days in Sind province, Karachi and Mirpurkhas. He left Karachi of the third of January and arrived at Bombay, India two days later. He wrote home reporting how busy he was with the work.

In Malaya the work seemed a failure due to my accident in Kuala Lumpur. Here I've not made any mistakes so far, although Clausen wants to blame the six days wait at Mexico City on me. He wrote a nice letter to Ted saying it was nice that the tracing of the shipment was so easy. In other words it didn't have to go way back to India into my lap.

Now, because I sent through to Mexico and it caused difficulties in New York he writes that they have had considerable trouble with my shipments there. He had a long telephone talk with Whitcomb about it (one shipment out of eight!) and said I should send them all through Whitcomb. Clausen has the thing all balled up but is apparently pleased to lay the blame over one or two shipments on me. One was six days in office in Mexico City. The other was delayed one day through poor routing by American Airlines.

The rest all came through okay. Clausen states that now he hopes my shipments will get through due to his arrangements with Whitcomb. It is just a change in name as my shipments have been going to his airport officer Mr. Post, and he has been transshipping them to Mexico. This was well arranged for when I was in New York.

I expected something like this from Clausen but expected it sooner. I saved the government $5 by shipping direct to Mexico as it goes anywhere on a minimum charge. It was a lightweight parcel. However, I didn't expect trouble at New York as the Pan Am line promised me they would turn everything over to Mr. Post for transshipment. Apparently they ship via another line (Eastern) which is slower and this delayed my one shipment.

Then I get a cable as though the world has blown-up and changes which are no good. I understand the cause but Clausen doesn't know "from a hole in the .ground." He is losing all my shipments which came through okay except for one held up in Mexico City six days. I expected blame for this shipment all along but wondered how I would get it. Apparently I left an opening by trying to save the government $5.

Anyway I was covered on that delay because all insect shipments go through Mr. Post and they were addressed to him. He apparently had trouble getting it from Pan Am. I knew all along I'd catch something if Clausen could possibly arrange it. But he can't make it up of course. The trouble is to keep from telling him where he gets off.

I didn't want to come out here in the first place. I've spent the most miserable time in my fifty years here. Although it might have been worse. At least I'm still alive and in apparently good health. What I'll get later I don't know. There is plenty of leprosy where I was. I was certainly run down with cold after cold. Anyway I keep my weight up and eat well.

I wrote Clausen the facts without mincing any words. All I care about now is doing my work here and I'll do it with no thanks to Clausen. He hasn't been help to me here. My reports have been way over his head or at least he hasn't studied them. I'll get all the parasites over from here that there are before I leave and I don't care what Clausen thinks. I certainly can't look for any credit from him even if I do succeed in getting a good one over and controlling the pest.

Fellow entomologist Ted Gardner had traveled from the Hoboken laboratory to Mexico to receive the shipments that Herb was sending via airplane. Gardner arrived in Mexico City on November 8th and set up shop in a corner of the laboratory that was operated by US Department of Agriculture scientists studying the Mexican fruit fly.

The fruit fly laboratory was established in 1928, and since 1931 had been run by A. C. Baker. The departmental annual report that year noted that "Baker has on various occasions expressed a wish to devote a major part of his time to research...and in the field of research has expressed a preference for the very important station which he had established in Mexico City. Inasmuch as the Mexico City station has recently lost several of its important research personnel, the time was particularly favorable to give Doctor Baker this opportunity." Whether Baker was being promoted or rather run out of Washington was a matter of gossip and speculation among staffers. Baker and the fruit fly lab were to become important elements in Smith's story during his work in Mexico.

It took two weeks to construct facilities that would be needed for the blackfly, including a quarantine room and a temperature and humidity control chamber. Gardner also traveled around looking for suitable citrus groves that were already infested with the blackfly and which could serve as testing and propagating sites for parasites that Smith was sending back from South Asia.

The records available for inspection today don't reveal how the earlier objection of the Department of Agriculture Pathology Department to air shipments was overcome, but apparently it was, as Smith began a series of air shipments.

Herb sent his first package containing 140 Catana beetles collected at Mirpurkhas on November 10th. The air express parcel left Karachi airport on the 11th, taking eleven days to make its way to Mexico City. The slow transit meant only 42 would be alive upon arrival.

A second larger attempt was made two weeks later. A large container with 2,500 pear leaves was estimated to hold 25,000 pupae of the orange spiny whitefly (*Aleurocanthus husaini),* closely related to the citrus blackfly (*A. woglumi*). There were also smaller vials, pill boxes and wooden dugout boxes containing *A. woglumi* pupae, along with 19 *Amitus* sp. and one *Eretmocerus* sp. The shipment left Lahore airport on November 23rd, arriving in Mexico City December 4th. No adult parasites survived the trip, though 55 *Prospaltella* sp. emerged from the pupae on the pear leaves.

On the 24th Smith wrote to Janet in Paris with news of his progress there.

> *I have been awfully busy the last two to three weeks and have had no time to write or anything. I have been hopping in and out of planes, trains, hotels*

and so forth almost daily. I carry my bed with me so I am okay everywhere.

I got rid of that fever finally. I still have colds as the weather goes from 35°F at night to 85° in the day. The rooms are chilly as there is no heat.

I sent some bugs from Karachi and a big box of stuff from here. These will give a lot of parasites, four kinds so far. I am having much more success this time – by using air shipments. My plants have almost all died but fortunately I can get more bugs here to infest them. I can't look after them when I am away and they don't care for them properly. I am leaving in a half an hour for Lyallpur. Then next week I will come back again for another shipment.

Karachi was warm and nice. Here I will start putting on long underwear. The rooms I live in are not very cheery. String beds – no springs. There is never a flush toilet. I ride in horse carts – open and awfully chilly except at noon. I wrap my blanket around my head like the natives. I have kept pretty healthy except for colds. Lots of malaria here, but with good food I don't get it.

I have a man who cooks my meals. I eat lots of curries and hot stuff. Bags of tea – three cups every breakfast – more at noon and sometimes social teas when I am invited out or in the villages. Then at 8:00 PM more tea. I still don't care for it. Every farmer I visit with has to trot out his tea. I'm treated like a big shot wherever I go into villages.

> *They pinch the stamps off letters here so you may not get all my mail. I now go to the special window and see the stamps canceled.*
>
> *The dust is truly terrible. On trains you choke with it. I may go to India next week. I will see the Taj Mahal if I have time. When I go to India Bombay will be warmer. On trains I often have the whole compartment to myself. No one travels first class much.*

Three days later Smith followed with a third shipment from materials collected at Gujranwala, Pakistan. Even though the package took a full two weeks to arrive in Mexico, Gardner was able to recover a number of live parasites: 36 *Prospaltella* sp. 5 *Eretmocerus* sp. and 13 *Amitus* sp.

Another large shipment was collected around Lyallpur at the beginning of December and shipped from Lahore on the third. This time the air transit only took three days and the survival of parasites was much improved. In one wooden dugout 22 adult *Prospaltella* sp. and 28 *Eretmocerus* sp. had emerged and survived. Over the two weeks following arrival an additional 198 *Eretmocerus* sp. and 961 *Prospaltella* sp. emerged alive.

Smith continued with more air shipments, sending a small wooden dugout of materials collected around Agra and Delhi by regular air mail leaving Delhi airport on December 7th. The fly pupae received in Mexico eight days later yielded a small of number of additional *Eretmocerus* and *Prospaltella* parasites. An additional dugout sent on December 9th failed to produce any live parasites. On the

14th another dugout left Delhi airport with fly pupae collected at Saharanpur and Dehra Dun, taking two weeks to make its way to Mexico City. One female parasite of a different *Prospaltella* sp. was recovered.

Herb made one final collection for the year, sending about 9,000 *A. husaini* pupae found around Lyallpur. The air express shipment took ten days, arriving in Mexico December 30th. A total of 105 adult *Prospaltella* sp. were recovered through the end of the year. The total haul from eight shipments was 1,076 *Prospaltella*, 258 *Eretmocerus*, and 15 *Amitus* parasites. Smith spent New Years Eve and saw in the new year 1950 alone in his hotel room in Karachi.

> *This is a hell of a place to spend the holiday. They don't do anything. I just stay in my room. Tomorrow, January first, I will go to work as usual. The hotels have celebrations but this always calls for evening dress and I don't have it. There are of course no "bars" as we have, to get a drink. It is quite different from any country I've been in, and I've been in a few. Anyway I expected this country to be as it is, knowing Morocco and Algeria and the other Moslem countries. They are all alike.*
>
> *If I could have made Bombay, as I tried to, I could have fared much better. India is better than Pakistan. When it was all one country under British rule, the Moslems had to accept the lowest jobs. The Indians, Hindus, got the higher ones. So centuries of this has bred up a low caste Moslem. There is only 5% of the*

people here who can sign their name and read and write. I have an awful job getting my voucher signed. Usually I don't get them signed. They just can't write. So you can see the condition of the country.

I am leaving January third by boat for Bombay. Things will be better. I mean civilization will be improved. I still think of Colima with longings after this. Poor food, poor everything. Fortunately I've not picked up anything as yet. Just ordinary skin diseases from not being able to take baths regularly.

Gardner has scared out a lot of my parasites and released them. No recoveries yet. Anyway I got them over. Will take a little time.

The only thing that keeps me happy is that I'm busy all the time. On the go traveling, sleeping in stations, in train compartments etc. If I slowed up and thought about it I'd be disgusted for what I've gone through. The natives sleep all over the floor, in the station, on the streets, anywhere. I usually have a string bed. I don't know many bugs I've picked up. I don't bother looking. I suppose my bed is lousy with bed bugs but they don't carry any diseases so I forget about them. There are too many diseases here to worry about bugs that don't cause any. Now I can get a bath occasionally.

The only thing that keeps these people from dying off like flies is the constant sunshine. It kills the bacteria. Flies are so thick here on meat that you can't see the food. And of course these flies visit toilets that are not flush ones. In this hotel they have plumbing fortunately. Usually a man goes around with a cart and collects the sewage. He is called a "sweeper" and is the lowest caste. The whole thing is too filthy to describe. I've got so I have to just pick flies out of my beer and off the food and go on drinking and eating.

They have the lousiest hours here. The business places start work at 10:30 and no one starts before 10 AM. How can a country do anything with those hours? They work about five to six hours a day. Yet the city is loaded with people sitting around doing nothing. Time means nothing here. To get a refund on a railroad ticket of $2.00 I had to spend one whole day, sign ten times and lord knows how many times they had to sign. They say it is the British System. "Too much paper work."

And no one checks it anyway! It will take a century or two for this country to get to where we were 100 years ago. Men who go to our Embassy for something are so surprised that it takes them time to get used to it. It's all blamed on the British.

Yet in Malaya the British were ruling and the system was good there. Boy, I appreciate what the British had to go through with in this country. Of course they had their own clerks, etc. They didn't mix with the natives and their pay was five to six times what they would get in England.

I'm the only real American in this hotel and of course I'm a bigshot. I would rather be in a decent hotel and just be a smallshot. But I don't like sharing

> *a room with anybody. I have too much work to do in the room.*
>
> *In the war it was different, I always shared a room but I had no reports to make out. Also I do my microscope work in my room. So I'm in this dump. Of course they are all surprised as no Englishman would ever stop here, and to them I'm an Englishman. There is a gang of kids about, the usual run of loose women, etc, but it won't be long before I'm gone. Anyway I'm the bigshot here and am certainly respected. I add lots to the hotel, apparently. In Bombay they say you get real hotels. Here I have a mattress on boards. I prefer strings and like sleeping on the floor. I only need a light blanket here.*

Gardner back in Mexico began taking some of the recovered South Asian predators and parasites and releasing them. In November 42 Catana predators were placed in a lime grove at Cuernavaca. Two weeks later 39 *Prospaltella* and 41 *Eretmocerus* were set out nearby. Gardner next traveled to Cuautla where 43 *Prospaltella* and 29 *Eretmocerus* were released among orange trees in a private garden. The releases continued most days over the next ten days in various orange, lemon and lime groves in Cuernavaca and Cuautla as parasites became available from Smith's shipments.

After a few days Smith left for Poona where he spent most of his time for the next four weeks. Several trips were made from Poona to distant regions in the province during this period. While in Bombay in late January, Herb sent another letter home.

> *I have been busier than the old cat on the tin roof ever since I arrived here four months ago. I should have an assistant.*
>
> *I didn't do much over Christmas and New Year. I just sat in my native hotel room with a few bottles of beer and stuck it out both times. Christmas I got up and went to the desert 200 miles away. Nothing like being on the go to make things pass by more quickly.*
>
> *Things are a lot better for me here than in Pakistan. I get decent rooms and feel like I'm out of the middle ages. I never even saw a woman in Pakistan for weeks, just those walking tents. They have "purdah," the women go around all covered up, with slats over their eyes. I didn't mind that so much as I did the accommodations. I've lived in worse places than our Negroes down south. Now when I have a room with a flush toilet I think I'm in luxury.*
>
> *The climate here is even better than Florida. In Poona it is 90°F during the day and 50° at nights. Bombay is 70° at night. In Pakistan I nearly froze my whiskers off, or rather my mustache. I haven't shaved my upper lip since September 18th.*
>
> *I've had cold after cold till I saw a doctor here and he gave me a dozen kinds of pills, one of which is to keep you from having colds for eight months, some British stuff.*

This is a top class hotel, full of Americans, and I feel pampered although there is no hot water and a few other non-essentials. It gets cold at night. Last night I started out with only a sheet as it was so hot. Then I got so cold I had to put on a sweater and a towel for a blanket and I shivered the whole night.

I think I've found the bug to solve the problem in Mexico. It controls it here. I have sent over rafts of stuff from Poona, which is near here. Also I sent a few others over just to help out a little. Only trials will tell the story. The trouble is if it really wipes out the pest I'm a marked man for future pest troubles in this line.

I'll be around Bombay and inland, middle India, Nagpur, for a month. Then I may fly up north for a shipment from various places. I may go on to Malaya and Sumatra depending how the parasites I shipped take hold. Gardner is down there in Mexico and handling the stuff. I had a lot of hard luck in air shipments. It took nine days from Karachi to New York. We don't know why yet. Most of the bugs were dead. I think they offload the stuff. But I've still got my plants, two tons of them. Two cages I keep with me everywhere, like a couple of suitcases but they weigh 300 pounds each.

Anyway life is picking up for me. I get a decent room, streets are fairly clean, fewer mud huts, better organization, and more English spoken. I felt like a tramp in Pakistan. I slept on station floors everywhere. Plenty of leprosy here but they say if you eat well and are healthy you don't get it.

Prohibition is my worst evil here. I get all set for a drink of beer and the shops are closed. After April first it will be dry all over. No one seems in favor of it. If I had to drink their water I'd get all kinds of diseases. But I'll be out of here by that time so don't have to worry about it.

In early February he returned to Bombay and left for Nagpur, Central Province, on the fourth. He wrote home from there that he failed to find the bug he was hoping for. The local entomologist had misidentified the pest. It was a different species of insect causing citrus disease there.

I bought you all a lot of silk and linen clothes here. Indian silk is $1.25 a yard. Linen is about the same. The Indians are the best tailors in the world. They do excellent work and cheap – a full dress shoulder to ankle – only $1.50 and I picked the styles from a fashion catalog. They make anything, any style. A full dress costs about $6.00 everything included, material (natural silk or linen) and the work.

I also got a sari (6 yards long and 46 inches wide). All Indian women wear a sari, no dresses – they just wrap it around themselves. I also bought a choli; it's a short sleeve blouse that shows the midriff. It makes a good sun suit. All Indian women wear this and put their sari over it.

Herb left on the ninth for Burhanpur, half way to Bombay. He returned to Bombay the morning of the 11th after an all-night ride. The weather had turned very hot - 90 to 95° all day. There was no rain and probably none until summer.

> *I didn't find my bugs in the last place either so only stayed one day. I am back in civilization again. I rode for miles in horse carts, bull carts, etc. There were no cars there and in many places no electricity.*
>
> *I haven't heard from Clausen in several weeks. Maybe he is in Mexico. The bug recently was discovered at the border, which will get everyone excited and there will be conferences. I haven't heard yet if any of my shipments from here came through. I sent nine. Gardner may be too busy but he should let me know as I am now leaving the country and going to Pakistan again. Too late to do anything here then. I'll be there a week and then I may go down to South India, Mysore etc. Then up again to the Himalayas and I'll be through.*

February 18 to 25 was spent in Bangalore, Mysore, and Coorg State. The next three days Smith surveyed Madras city and the surrounding area. March 2 to 6 he was back in Bombay, then in Delhi four days, then a side trip to Saharanpur for a few days. Returning again to Delhi, Herb flew over to Calcutta on March 11.

It is very hot and sticky here. I leave tomorrow at 6AM for way up near Tibet by plane. After that I will have covered all of India. They have the most rain fall in the world there – 500 inches a year – and Rutherford has about 30 inches! I have no rain coat with me either.

Ca1cutta is lots bigger than Bombay. Hotels are not crowded and I had no trouble getting a nice room. My bugs were not ready to ship. I didn't want to loaf for ten days so I came here.

This is one of the old swanky English hotels. Swell dining room etc. The English sure lived like kings here. No wonder they hated to give it up.

Where I am going is like Malaya and I'm hoping to find the bugs from Malaya there. I have found only three of the five good ones. If I find the other two I'll head for home and not have to go to Malaya.

I have luggage straggling around all over India. A steel trunk is in the Bombay Cloak Room at the station. My old trunk is in the station in Delhi. Two cartons are at the US Embassy, Delhi and my typewriter is in a repair shop there also. I hope I remember to pick them all up.

I left Bangalore and forgot to pick up the silk I bought you. They sent it to me registered. Everything has to be locked here. They won't take it in the Cloakroom otherwise. I haven't lost anything yet.

I had to stay in a real native hotel in Delhi. It is the worst place to get a room in all of India! The Embassy cannot even do it. No one goes there unless he

> *gets a reservation three to four weeks in advance, and I can't do that.*
>
> *I went by train to Delhi. I had a four-berth compartment all to myself, with bath, toilet etc. The only white man! Every one of our class uses plane. You just lay your bed roll on the bench and go to sleep. I don't think there are any bugs in it yet. I left my bed roll behind this trip – another place to look-up when I return to Delhi, the native hotel. You should see some of these places I have to sleep at. I sleep on station floors with hundreds of natives around me. They put a piece of cloth down on the stone floor and curl-up and go to sleep. Every station has hundreds of men and women sleeping on the floor. The food is filled with chilies (red peppers) No decent food here, but I have felt OK, better than in Mexico.*

On March 12 Smith flew to the northeast states spending a few days each at Gauhati, Jorhat, Shillong, and Gauhati again. From there it was back to Calcutta for several days and then on to Delhi in March 21st. Smith then spent the next ten days traveling around he Delhi region with side trips to Saharanpur, Bareilly and Agra.

By the spring of 1950 Smith had been traveling almost constantly for seven months, usually working seven days a week, often sleeping on a bedroll in train stations and exploring rural byways by oxcart. His last three air shipments were sent from India in April, containing in total 160,000 pupae of *Aleurocanthus woglumi*, as well as

6,000 pupae from related species and several hundred collected parasites. Over several weeks after arrival in Mexico, 1,997 *Prospaltella clypealis* and 1,939 *Amitus hesperidum* parasites emerged from the blackfly pupae, along with smaller numbers of other *Prospaltella* species.

Smith was largely operating on his own devices, as he only received occasional communications and little useful advice from colleagues back at the Department of Agriculture headquarters.

> *Clausen made a lot of mistakes on his trip to Malaya and my work is showing it up. He got the parasite importance all mixed up. I have asked him his opinions but he never replies. If I write a report it is going to make his work look pretty foolish in some ways. When he was here he just hit the high spots. He wouldn't bother about using a high power microscope to determine the parasite as I have. In those days they didn't bother, but he hates to admit anything. If he asks for a report, a big published government bulletin, I'm going to tell him it will show up his work in Malaya. I'm not afraid to as I can retire any minute. When I think of the energy I've put into this job at 55, I can get much more out of private work, probably on a farm or something like that.*
>
> *I have put in seven days a week here and nights on this job. Not because of any credit I may get, but mainly because I would go nuts if I had a day's vacation. Nothing to do, nothing to see, no bars, no*

drinks! You are only happy if you are working. No other country is like this.

Herb had reserved a space on the S. S. Exchequer which was scheduled to sail on or around May 2nd. His work in Pakistan and India was done. It was a difficult seven months, but his results were extremely promising. He had succeeded in all he had planned to do in the region. Herb reflected on his experiences in South Asia in a letter to his daughter Janet.

In India, cholera, smallpox, bubonic plague and minor diseases were as common as measles in U.S. and one pays about as much attention to them unless he happens to read that part of the paper and sees, 300 people died the past week from cholera, etc. After that you just don't glance at that part of the paper. I had to go three times to the source of the Ganges at Hardwar where millions come from all over India every 12 years to bathe in the river. They always had a terrible epidemic there in the past, and brought it back to their villages.

This year, the Indian doctors inoculated over a million. The fact is they would not allow my assistant and myself to buy a rail ticket to there until we could produce a certificate of inoculation less than a week old. They were sleeping all over the station floor, streets, just everywhere except inside. They have no money. Children, women, everybody. I have also slept on railway station floors, but I had a bed roll under me.

People do not realize that India has gone back in many ways since the British left. No one worries about these diseases though. Europeans rarely get them. We eat so much better than they do. They do not even have soap to wash their clothes. They pound them on a rock in the rivers, swinging them up and down.

Smith was able to embark on the S.S. Exchequer, which steamed out of Bombay on May 15th. The ship called at Djibouti, where Smith was happy to speak French again in this former colony of French Somaliland. It brought back old memories. Writing to his eldest daughter in Paris, Smith wrote that Djibouti had "real French bread, cafes with Pernod, St. Rafael and Suze."

It was intriguing to look into the little restaurants and see a small piece of French bread in a napkin in each place. I had to hustle back on board as I bought a case of beer and might have forgotten them if I had stopped to sample the French life with a few wines. At Port Sudan – that is British – no life but very correct and well organized – too well. I like the French organization better.

When the ship arrived at Port Said, the passengers and crew were not allowed to disembark, as Djibouti was in the midst of a cholera epidemic, and the Exchequer had to fly a yellow quarantine flag. After spending a half a year in South Asia, Smith didn't worry much about diseases.

As the ship neared the American coast and passed the Grand Banks, Smith wired for flowers and sent a radiogram to his wife and youngest daughter Lucienne who were on a ship steaming nearby towards Europe.

Whiling away the long hours on ship, Smith composed a modern update of an old Kipling poem:

I came home from India on a freighter,
Along with cages of citrus
One night they all died
When I said: "I wished I could throw them over the side."
And I learned about cages from these.

He sent the poem to his daughter Janet in France, asking her to pass it along to Harry Parker, who had since returned to Paris. He added that "Dr. Parker knows these old college ukulele ballads much better; I learned most of mine from him, after we got down from the mountains and relaxed."

Smith's travels around the Indian subcontinent had been long and arduous. The collection of old-fashioned Wardian cages weighing over a ton that he had dragged around via trains, trucks and oxcarts proved to be more of burden than a help. He had shown with this expedition that air shipment of parasites was the way of the future.

10

Mexico: Insect Wars

The Exchequer arrived in New York on June 9, 1950. After a quick dash to drop off souvenirs at the family home in Rutherford, New Jersey, Smith was ready to head for Mexico. Passenger space on regular passenger ships going from New York to Vera Cruz, Mexico, was unavailable before August but the AGUI Line had a small freighter, the S.S. Coastal Delegate, sailing June 22, and passage was obtained on this. Smith spent a few days at the Washington headquarters and caught the ship at Baltimore, sailing from there on June 25th. The ship was expected to arrive at Vera Cruz July 4th, but Smith got off instead at Tampico because Vera Cruz was in the uninfested zone.

On July 7th the ship pulled into Tampico harbor. Smith immediately got to work. He studied parasite colonization sites in the Valles region for two weeks before arriving in Mexico City on July 20th. The rest of the summer Smith was busy with frequent field trips to Valles, Guadalajara, Colima, Los Mochis and Mazatlan.

At the beginning of October Smith took a rare break from field work to give a blackfly tour around the Valles area to a group of scientists from the Latin-American Congress. Quickly back to work, Herb traveled around the colonization sites, with a special focus on the Cuernavaca and Valles sites.

Between August and December more than 23,000 *P. smithi* parasites were collected from the tree where they were originally released. In a three month period 6,000 parasites were released at four sites around the Valles area. Another 700 were released at Los Mochis, along with 500 at Guadalajara and 200 around Cuernavaca. By the end of the year the parasite colonies were well established everywhere they had been set out around Valles. While no more than 5% of the blackfly pests were parasitized, Smith's earlier results in Cuernavaca indicated they would build up over time to attack most of the blackflies.

Back in Cuernavaca, adult parasites that were found in abundance in October had largely disappeared by December. This was because the citrus blackfly had entered its third stage of life. Blackflies in the pupal stage found in nearby trees became the next targets for parasitization with *P. smithi*.

Several other species of parasites were not faring so well. Colonies of *P. clypealis* were struggling, with only scattered adults found where they had been released at Valles and Cuernavaca and no offspring were seen. The situation was similar for the *Amitus hesperidum* parasites

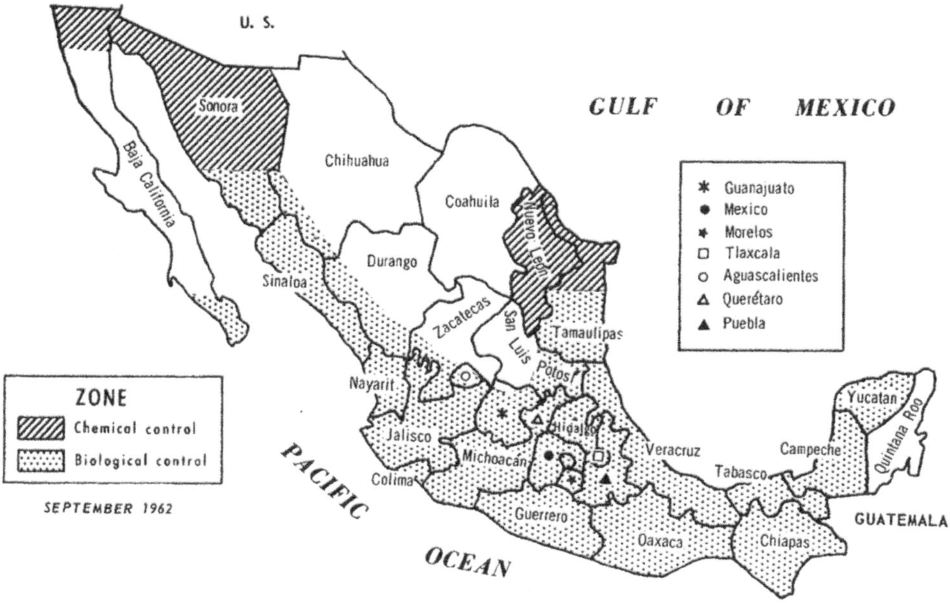

The Americans favored insecticides over biological control. Smith was allowed to work with parasites in Mexico with the restriction that the Mexican states bordering the United States were to be "chemical control" areas where the blackfly would be sprayed.

released around Cuautla and Cuernavaca. *Eretmocerus serius,* which had achieved a parasitization rate of 85%, dramatically dropped to 5% during the dry season. Smith found that the dry air made the honeydew become sticky and the emerging adults became trapped. *P. smithi,* being an internal parasite, was unaffected by the low humidity.

Smith spent much of his time studying the biology of the citrus blackfly along with the parasites, especially *P. smithi*. At Valles, the blackfly had three to four generations per year, overwintering in one stage from October to February.

While Smith's first interactions with the Americans at the Mexican Fruit Fly Laboratory in Mexico City were cordial, tensions developed over the direction and control of the blackfly effort. Such problems were really nothing new, as within the large bureaucracy of the U. S. Department of Agriculture there were frequent skirmishes between different sections about who was in charge of programs. Smith's section was usually at a disadvantage as other sections claimed territorial control and the parasite men had to ask for permission to do their work.

The head of the Fruit Fly Lab, entomologist A. C. Baker, was a well entrenched veteran of the department. He had been studying the Mexican Fruit Fly there for more than 15 years. It was a big budget operation with over a dozen entomologists and supporting staffers. They did good science in the classical sense. There were well-designed experiments, measurements, statistical analyses, reports, published papers, conferences and so forth.

The problem with all this from Smith's perspective was that the Mexican fruit fly problem was worse than ever. The other problem was that they wanted control of the blackfly project. Smith clearly wasn't motivated by a desire for power or acclaim. He had no interest in moving up the hierarchy to an administrative desk job. He wasn't

Herb traveled hundreds of thousands of miles around Mexico working in the field and meeting with local agronomists and farmers. He was often accompanied by Mexican government entomologists or agronomists, but his Spanish was good enough to go out on his own, and he often went weeks at a time without speaking any English. (Smith family collection).

much interested in conferences and papers unless they furthered getting the job done.

Herb didn't expect to be recognized at all for the work he was doing. But he was damned if after all of the long-hours of hardship and misery out in the field away from home, he was going to let others waylay the project and stop him from finishing the job he had started. The job in Smith's mind was nothing other than to get the blackfly under economic control as soon as possible.

By January, 1951, there was interest in studying parasites in Texas in spite of the general ban on stateside parasite work. Headquarters didn't want Herb to bring his blackfly parasites with him, only to study what other beneficial parasites might be killed off by the general pesticide spraying they planned for the groves in the area.

Smith wrote to his family: "I will probably have to work in Texas near Brownsville for a year studying parasites, as they are going to spray there once the blackfly gets in and this will upset the natural control by parasites of their other pests." Texas citrus growers, unlike those in California, had not been using chemical sprays and had been relying on natural parasites to control a variety of pests. Smith had already been traveling widely throughout much of Mexico, and this new plan added to his difficulties. "I will have to work back and forth from Texas to here."

For most of January 1951 Smith was working in Cuernavaca collecting *Prospaltella smithi* parasites and shipping them to various regions. In early February Smith traveled along the west coast surveying the parasite colonies and releasing more parasites in Navojoa, Los Mochis, Mazatlan and Guadalajara. While in Los Mochis Herb wrote home, reporting that one parasite was becoming well established here and was doing better and better in various parts of Mexico.

> *Looks like I 'm going to be down this way for another year. I am supposed to run back and forth from the Rio Grande Valley to Mexico. In the Valley I have to study the natural enemies of all citrus insects there. This has never been done as yet. They want to have this information now, as sooner or later this black fly of citrus will get up there. Then they will start spraying with vengeance to eradicate it. This will upset the nice natural control they now have.*
>
> *The Mexicans don't have to spray their trees as California does since the parasites keep the pests down. Insecticides will kill the parasites and then those pests will loom up.*

Curtis Clausen retired from government service January 31st. He spent a few months golfing and relaxing, but then the University of California Riverside offered him the chair of the Biological Control Department there, which he began in July. Smith had good relations with the entomologists there, especially Stanley Flanders. He discussed parasite issues with them from time to time.

> *Gardner is more or less taking his place. He is second to a Mr. Latta but will handle all the parasite*

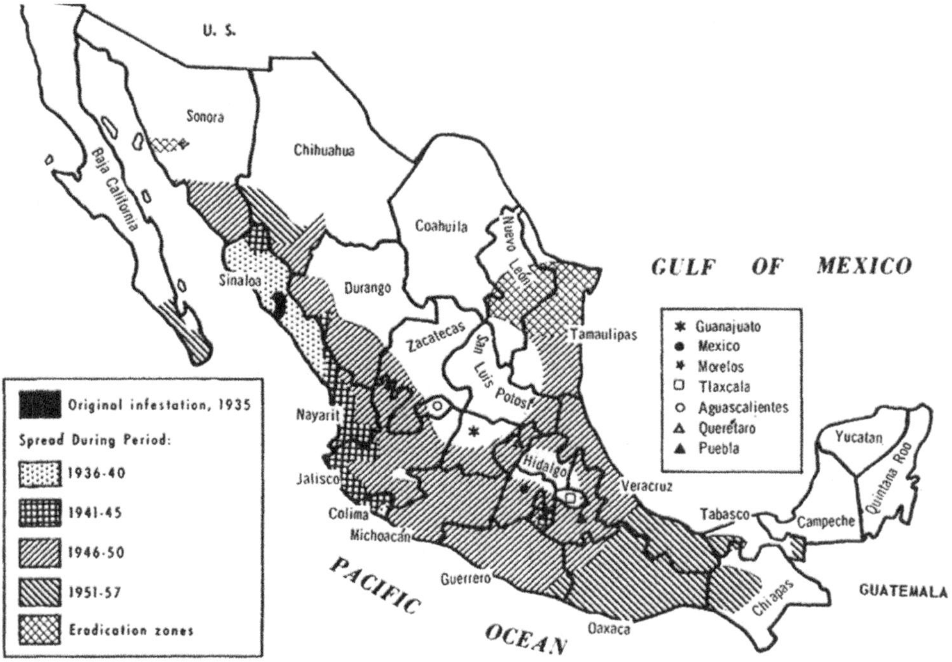

After seeing the initial results with parasites that Smith had brought back from Asia, the Mexican government was enthusiastically supportive of Smith and the parasite work. Smith worked with the Mexican government agriculturalists to spread parasites throughout most of Mexico.

matters. Latta is only an insecticide man and knows nothing about parasites. Gardner has always honeyed up to Clausen so I'm not surprised about his new position. Gardner has very little grasp of parasites but is good at regulations -- fussy over these things, red tape etc. I just shoulder that aside and get after the research part, get out into the fields and use a microscope.

They couldn't have put me in (even if they planned to) as I have to be on the project. I have too much experience and background on this particular work to put a "green" man in. When I was with Gardner I always did the parasite research work and he sat in the office and did the office work. He didn't do too well in getting my parasites going here but we had no one any better. I wrote him all the instructions when he was here, what to do and how to do it and so on. Even then he slipped up on certain things.

Smith had finished his work in the Los Mochis area and had been stuck several days with nothing to do waiting for a flight out. He went to the airport every day but the planes were always full.

The first two days, all hotels were filled and I had to sleep at a sort of brothel. The girls didn't bother me though. The fact is they were highly amused at my putting up there.

After getting bumped off the plane yesterday I went back to my "room" and drank some boiler makers. My tongue really loosened up and I heard about it this morning. Apparently I made a big hit. I wish I could remember what I said to them. I can really throw the Spanish around now, but with a little liquor I really pour it on. Anyway I woke up safe and sound in my own bed, alone, this morning.

> *I got bumped off the plane again today as some guy gave them ten dollars and got on. Was I mad! Anyway I have a decent hotel room now, probably the one belonging to that guy on the plane.*

In late February Smith was working in Valles, and after that was in the capital for a while. There was a short trip back up in Weslaco, surveying parasites in the lower Rio Grande Valley. For most of April and May he was at Valles, with the exception of one short trip Sonora and northern Sinaloa. From thence Herb went to Guadalajara, Colima, La Paz in Baja California, and back to Valles.

That spring the entomologist P. A. Berry was hired to make yet another search for the elusive Catana. While Smith had successfully brought back a half dozen parasite species that attack the blackfly, the varied and changing results seen with those species suggested that in the parasite field more is always better. It was impossible to predict what would work best, especially in a large country like Mexico with many climatic zones. After nearly a month of work in three provinces, Catana was still nowhere to be found.

On June 12, 1951, Herb flew back for meetings in Washington followed by a visit home at Rutherford. His daughter Lucienne was out of school for the summer, and Herb and his wife Jean decided the three of them would spend the summer in Mexico. It had been several years since Herb was able to spend more than a few days at a time with his family. Since his work in Mexico was likely to continue for quite a while, this would give them a chance to see how life in Mexico might work out. They drove through the American Midwest in their new Oldsmobile, arriving in Mexico City on the 30th.

The difficulties soon became apparent. Herb's work required to spend most of his time in the field, criss-crossing from one end of the country to the other. He had an apartment in Mexico City but he wasn't able to spend much time there. Sometimes when Herb would have to be out in the field all day, the ladies would visit with friends they had in Cuernavaca. The trio were able to travel around and sightsee that summer, staying in nicer hotels than the rough-and-tumble places Herb would usually stay in on the road.

Between tourist stops Herb would make the rounds of various citrus groves that he was monitoring. Lucienne recalled how comfortable her father was visiting and chatting in Spanish with the workers and the local people and how warmly they greeted him. He would lift the pot lids to see what they were cooking and they would all joke with each other.

In one village Herb mentioned Lucienne liked riding horses, so someone brought out a horse and everyone in the town came to watch her ride. Unfortunately the horse didn't seem to understand her Spanish, as it trotted out of town and then refused her command to return until a farmer in a field gave her a switch to get him to go.

Lucienne noticed the other American entomologists at the lab in Mexico City never socialized with Herb, nor did they work together at all. A Spanish lady at the lab named Mrs. DeWit was aware of the tense situation and helped Herb get what he needed from the lab. At the end of the summer, Jean and Lucienne hired a driver and returned home to New Jersey.

In the third quarter of 1951, Smith reported that he had made four trips to the Valles area including Tampico and many trips to Cuernavaca and Cuautla. In August he traveled around from Guadalajara and Colima, and then spent eight days near the coast in the State of Guerrero. In September he was in Ciudad Obregon, Navojoa and Los Mochis.

A severe hurricane that hit the Valles region in August set things back there, as the parasites seemed to suffer greater losses than the blackfly. The *Eretmocerus* parasites there usually increased over the rainy summer months, but after the hurricane parasitization rates decreased from the typical 70% down to 30%.

In early October Smith was working back in Mexico City and sent home another update on his activities.

> *I have been pretty busy the past few months checking on the parasites released. I had to cover practically all of Mexico. I even had to take the slow train on the west coast. The train runs about 20 miles per hour because of poor roadbed.*
>
> *One parasite seems to be doing well, but it is one I least expected to do anything. Two others are established but have not built up much yet. These two were the most important ones in India. Releases were not just what I would have liked but Gardner did the*

best he could considering he knew very little about this work.

From only 126 adult parasites on one tree I have collected about 30,000 to ship all over Mexico. Yet the other parasites had 20,000 released (from India) and I could not collect more than a couple hundred. Anyway parasites are peculiar and they may build up more later on. This third one really goes to town on the blackfly but I won't know its value for a year or so. I don't know yet how long I'll be here.

There have been lots of Americans here the past few months. Hotels are always full. It is pretty nice in some places such as Cuernavaca and Acapulco. Food is excellent and reasonable. I eat at least one filet mignon a day. It costs about 50 cents in a decent Mexican place (but not for tourists). Hotels are modern and reasonable, $2 – $4 per person. The whole country is going "American" and going fast. There are lots of fine new buildings, just as in New York City.

They will never change their traffic customs though. You hear a lot about the traffic here but it's easy to drive around after you get over the first day. No one hits you but they seem to try. Cops don't bother tourists no matter what they do. Also I have "U.S. Government" on my license plates and that helps.

We are no longer "Gringos" here, although the politicians think it lots of times. If anyone wants to sell a new product they just tack "American" on it. I hear loud speakers selling things, always something about American in it so must be good. '

A ride in a two cent bus is really something the first time. Tourists usually take a taxi. The climate is okay in Mexico City once you get used to it. It is always sunny, with lots of flowers. Carnations are ten cents for two to three dozen on the highway.

Every corner has a man squeezing orange juice costing five cents for a pint glass. Strawberries have been in the market here for a month. I tried some "pulche" (a drink made from juice of agave plant), a poor man's drink. For five cents you get a real jag on, if it's good stuff. Good for the stomach too, if you don't drink too much. '

The last week of October 1951 Smith was in the Lower Rio Grande Valley studying the natural enemies of the citrus pest there. The only pest there causing any damage was the California red scale, and that was usually seen in groves of young trees inter-planted with cotton and dusted with insecticides. He returned again in early November and in mid-December. During that same period there were frequent trips to Cuautla and Cuernavaca. Smith was covering a lot of ground.

A formal commission known as the Mexican National Committee for the Control of the Citrus Blackfly was created in September by a decree of the President of Mexico. Manuel Moreno Torres was appointed as President of the Committee. What had been a patchwork of State and Regional Committees were now brought together.

Smith felt this arrangement greatly facilitated the work. A Biological Section with around ten men operated out of Mexico City, receiving and shipping or hand-carrying the parasites to the various regions. Most areas had their own men to receive and release the parasites. At the end of 1951 Valles, Cuernavaca and Colima all had their own parasite collectors. In January 1952 another crew was to start up in Villa Juarez to the west of Tuxpan, Veracruz. The plans were to add more local crews as the parasites built up in other regions. The general overall handling of the parasite collecting and colonization work continued to be under Smith's direct supervision.

The Northeast Committee sent a large group of inspectors to survey Tampico port and surrounding areas and identified 6,000 properties with infestations. In October and November, over 200,000 parasites were released in the area.

On May 19th, Smith wrote from Mexico City to his daughter Janet in Paris:

> *The Mexican government is all for parasites now so I am pretty busy running around. They have men in all regions collecting and spreading around the bugs. They have practically stopped all spraying. Looks like I'll be down here at least another year. The Mexicans don't want me to leave. They say they can't handle it themselves. There is still a lot of research to be done on the parasites and they have no one to do it. I work more in the Mexican government offices than I do at our Bureau lab.*

On August 12· 1952 Smith wrote to headquarters asking them to arrange a trip for the Bureau chief Hoyt to visit the work in Mexico:

> *Probably one of the most important reasons for Mr. Hoyt to visit Mexico and look over the blackfly situation is the growing importance and effectiveness of the National Committee for the Combat and Control of the Citrus Blackfly in Mexico. This organization has gone all out for biological control since its inception last September. It was set up at the instigation of President Aleman because of the failure of the State Committees to control or eradicate the pest. Also only four states showed any interest in combating or trying to control the pest. In order to have combat and control nationwide the President laid out plans for the above Committee which was to be directly accountable to the Secretary of Agriculture.*
>
> *The National Committee has now stopped spraying everywhere except in the Monterrey – Victoria region, where they are doing eradication work. This leaves the spraying entirely up to the individual growers. In the southern half of Mexico very few growers can afford to spray and those that can lack the know-how to do a good job.*
>
> *As a result there is considerable confusion as to just how far one should depend on parasite control and how much dependence should be placed on spraying until the parasites have given control in more regions. The National Committee has been depending*

entirely on parasites everywhere even where the infestation is high. In the Valles region they have refused permission to three growers to spray their own heavily infested groves and are now trying to saturate it with parasites.

If the Committee has such strong authority they could also stop Mr. Cooper's spraying there. It is the Committee's understanding that there should be no spraying where biological control is carried on. I have recommended spraying the heavily infested groves and then releasing parasites on the resulting low population. But apparently, now that the Committee embarked on an out-and-out biological control program they fear that they cannot alter it any because of possibly psychological reasons.

Mr. Hoyt's opinions on the chemical vs the biological control aspects of the blackfly would carry much more weight with the National Committee than that of either Dr. Baker or myself. As a result, there would be a much clearer understanding of what is to be done in the future.

Between October 8 and 20 would be a very good time to inspect the blackfly work here, either from a chemical or a biological control angle. It is the end of the rainy season when normally the blackfly is at its peak. Thus a comparison could be made with the groves clean due to parasites and those with little or no parasites.

The Chief could complete his program here in from 5 to 10 days, with a five day visit he could meet

> *the National Committee officials and the Secretary of Agriculture (1 day), inspect the biological control work in Cuautla and Cuernavaca (1 day) , conferences in U. S. Laboratory and Mexican Official Reception (1 day), and a trip to Valles (2 days) . If more time were available a trip to Sonora could be included (4 days).*

At the end of 1952, Smith wrote a Christmas greeting to daughter Janet in Paris, noting that he was on his way home to Rutherford by government car, the first trip home in four years. He mentioned his brother Lee had retired after selling his Ohio bicycle factory and was planning to spend winters relaxing in Florida. Herb on the other hand had a harder life, traveling thousands of miles a month around Mexico, out in the field, with only rare visits to his family. But with all his activity Smith noted "people say I am getting younger." Most importantly the parasite work in Mexico was progressing well.

> *It will be the biggest parasite success in years. The pest is wiped out in several states. The Mexican government is very pleased. The insecticide people and the men at the lab are not happy. I don't go to the American lab except at night. I work in the Mexican government office. I speak Spanish better than the men at the lab who took lessons and have been here 14 years. I travel everywhere with Mexican officials and go two or three weeks without a word of English.*

Upon his return to Mexico in mid-February 1953, Smith wrote at length to his boss James Hambleton back at the Washington headquarters advising him on the situation there. He reported he had visited the offices of the new National Committee as soon as he arrived in Mexico City. Smith diplomatically wrote that he didn't go to American lab that "of course was closed otherwise I would have gone there first." No mention was made about the poor relations the lab men had with Smith and with the Mexicans.

> *The reception by Agr. Escalante* and his Secretary Agr. Monigo Rios Rosales was exceptionally cordial. They gave me the impression that I was the most important one in the Committee, probably because of their strong conviction in biological control. Any and all of my instructions or advice will be carried out, they said. So far they have done this.*
>
> *I am definitely pleased with the new set-up and feel that I can do an even better job than I did with the old Committee. Outside of the President and Secretary there has been little change in personnel at least in Mexico City. The men of the previous Committee in general, are working just as enthusiastically for the new Committee with no friction except in one or two cases.*

---

* In Mexico, scientists with a degree in agronomy take the title Agrónomo, abbreviated Agr.

The main thing is that Escalante, his Secretary and all the rest are behind me in the work. They still feel that they are a long way off from taking over themselves, at the present time. I feel so too. They lost about 38,000,000 Amitus somehow in Michoacán. If I had been here this would not have happened. They used their own ideas instead of my instructions which was part of it. Sometimes they put the bags containing Amitus out in the field without even opening them. Amitus cannot fight its way out of a paper bag.

For a while now my work will be in great part administrative, getting the new administration organized and functioning properly and I do not mean only parasites. They have placed an awful lot of confidence in me and I do not like to let them down by just sticking to parasite work. There is friction here between some of the old and new Committee men and this is going to require considerable diplomacy and tact to smooth out.

Smith saw his job had become part scientist and part diplomat. He felt he should work wholeheartedly with the new committee, even though some of the men on the old Committee were resistant. Smith was willing to work with anyone wanting to help.

They think Escalante will not last long. I intend to play ball with Escalante and his men so that he knows he has absolute confidence in me. So far he does have that.

The Mexican Biological Control Department moved into labs in the 20-story Mariscala Building near the Belles Artes and the Alameda Central park. When not out in the field Smith worked out of this lab rather than the American lab across town. The Mariscala Building was damaged in the devastating 1985 earthquake, and was condemned and demolished.

At the same time I have to be careful not to antagonize these old Committee men. Ing. Jimenez is one of these latter. Incidentally he suggested that I give the new Committee little information on the work as they will use it and then take all the credit for it. I do not believe this and will continue to give all the help I am capable of to bring a successful conclusion to the work, no matter who heads up the Committee.

The Mexican agronomists went by the title of "Ingeniero" meaning engineer, a shortened version of "ingeniero agronomo" and reflecting the concept of scientific agriculture as a type of civil engineering. The title is usually abbreviated "Ing." Ingeniero Eleazar Jimenez Jimenez worked closely with Smith and became one of the leading figures in the Mexican biological control field.

Smith's Mexican colleagues in turn often referred to him as "El doctor Smith," though Herb always made it clear he did not hold a doctoral degree. This was typical for most entomologists of the day, even many of those teaching at universities, who generally opted for getting out into the field and learning on the job rather than spend the time to complete a Ph.D. in Entomology.

My job down here is turning into something bigger than I anticipated. I have to smooth out these frictions and get all the men to work together as one team and there are at least 200 of them all over Mexico. Jimenez says that if he leaves he can get a good job under Ing. Moreno Torres, the head of Public Works of Mexico City and a much bigger job than President of the Committee with lots of money and personnel. He says twenty or more key men all over Mexico will walk out with him. This would set the work back six months or more.

I have the confidence of all these men and of course know them personally and trained them. That is why I have to get around Mexico as soon as possible. Moreno Torres took over all the fired old Committee men. I visited Moreno Torres soon after I arrived. He wrote me a letter asking me to continue with my cooperation as before with the new Committee for the good of Mexico. He is very powerful here. He wanted to give me a big dinner with the old crowd but I was able to duck it because of another appointment. In his letter he expressly asked to see me when I returned.

I have established myself with the Mexicans as unbiased and only want to do a good job. They like me for my close cooperation and informality, unusual among Americans here. These are their own words. Things have slipped considerably during my two months in the States, most of it probably due to the change in administration. The only region that worked well was Valles where my assistant Zavala continued as before. He is my best man here. He is unaffected by personalities like some of the other men, many of whom act like primadonas.

Smith spent all of previous week in the field around Morelos, Guerrero and nearby States. He reported the pressure of the work prevented visits to the Americans at the U.S. laboratory. He said he could only arrange to be in the city over the weekend, which conveniently was when the Americans were away from the lab. He did allow to Hambleton that he did hope to have some time to see Baker and the other men soon.

> *The Committee wants me to go immediately to Veracruz to reorganize the work there and advise on a new outbreak, recently discovered south of the port about 50 miles. Also I must check all the other places. Most of the men in Veracruz are new, which is natural as it is Escalante's State.*
>
> *This job is really too big for one man but I have the bull by the tail and if I let go the whole thing may collapse. However, if I do not have to do any detailed research work on parasite-host relationship in the next few months I can easily swing it.*

Herb was pleased that the Committee Secretary planned at Escalante's suggestion to travel around with him to all regions of Mexico to check on the parasite work and even spraying. Herb found the old committee would only send Jimenez with him, but Jimenez had little political authority compared with the Secretary.

> *Agr. Escalante gave me a desk (new, modern, steel much better than my kitchen table previously used), filing cabinets, etc. in one of his offices, and I*

will now spend all my time there when not in the field. They were very pleased at my acceptance.

My work cannot be done by remote control from the U. S. laboratory. They have 25 or more men daily in the biological laboratory, many of them new, from the various States for instruction. I definitely have to see that they start right and the Committee wants it that way. These men take much more stock in what I say than in some Mexican who they know never studied entomology.

As far as the language goes, they get it, never have to repeat myself. I have now been told that I speak Spanish better than even Stone at the laboratory, who was the best one there. They like my Latin accent better than the American. (This is due to my French and Italian which I still speak a little better). Incidentally, my rest [from speaking] Spanish while in the States worked wonders. The Committee noticed it. This means that if you want to speak a language well take a vacation after you think you have it down well and then return. I even noticed how much more fluently I spoke myself. It just poured out).

Escalante offered to buy Smith a new car for his work, and Herb asked Hambleton if that would be acceptable, as the old Ford the Americans had provided him was constantly breaking down as he traveled around remote areas.

Also Smith wrote headquarters that they should start corresponding with him through the Mexican offices in the Mariscala building since he was usually there and if he was out in the field the Mexicans would know where to find him. The Mexican office would forward all air mail letters and telegrams to Smith in the field.

> *Also we keep constant telephonic correspondence with the central office at no expense to me. This is a decided advantage in my work as I have to confer with the Central Committee on problems and on the same day that they come up. In this way we get prompt action. Incidentally this is rather new in Mexico and they are glad to have the opportunity to be taught American methods.*
>
> *This will be all along this line at the moment although I have plenty more information. Being in the field so much my research work, checking on the parasites, dissections, etc., gets behind. This work in a way is more important than letters just now as things are changing fast, and to the better, as far as control of the blackfly is concerned. Valles is even cleaner than before. Zavala is putting out traps almost everywhere as he cannot find the blackfly living forms by visual means.*

On April 15, 1953 the Mexican blackfly committee wrote to Avery Hoyt, the Chief of the Bureau of Entomology and Plant Quarantine:

> *I am writing this letter by the specific orders of O. Agr. Celerino Escalante E., President of the Na-*

tional Committee for the Combat and Control of the Citrus B1ackfly, in order to thank you for the cooperation that your Government, through your office , has given us in the Campaign against the b1ackfly, commissioning Dr. Herbert D. Smith to us.

It is a great pleasure to us to inform you that we have obtained very satisfactory results in the biological control of the plague and that Dr. Smith in this respect has been an untiring and wonderful collaborator.

With the firm conviction that our governments will continue giving each other a mutual cooperation that will reflect itself in the progress and greater closeness of our countries, I remain,

Very truly yours,
Secretary of the National Committee

Around the same time Harry Parker also wrote to Hambleton to commend the success of the blackfly project in Mexico and to encourage the department to let Smith continue without undue restraint. Parker by this time had returned from Uruguay to man the old European Parasite Laboratory outside of Paris. He wrote that "Smith is doing a fine job in Mexico and I hope nothing will interfere with his continued good work there."

Parker continued that projects like the one Smith was carrying out "have a great effect on international relations, often arousing more sympathy and good will towards us than is created by people high up on the Ambassadorial shelf." He pointed out the problems of successfully working in a foreign country. "I get some idea of Smith's difficulties in his relations with the various factions, and between the lines one gets a distinct whiff of the old *sui generis español.*"

Parker also noted the importance of Smith's linguistic skills. "I am not at all surprised that Smith is doing well both in the work and in the language. He handled French and Italian well when he was in Europe." He added that he knew a quick test to see if someone could handle agricultural French. "Translate *Yee* and *Haw* into French – it is *Hue* and *Dia*. Well Smith knows *Hue* from *Dia* all right."

Tensions remained high between the Baker and the Fruit Fly Lab on one side and Smith and the Mexican Blackfly Committee on the other. In July L. F. Curl, the head of the Department of Agriculture station in San Antonio, Texas, wrote to Baker mentioning he had read in a report by P. A. Berry that the head of the blackfly committee Escalante was "not too friendly with the Americans." Baker evidently took this personally as he responded with an angry memo stating that Escalante "has always been very friendly to me and I look on him as one of our good friends.... Of course, there are some Americans I have bumped into in Mexico whom I would not want to be on particularly friendly terms with myself." The latter jibe likely was a veiled reference to Herb Smith.

Smith continued pressing forward with the blackfly parasite work together with the Mexican blackfly committee. A total of 1,600 workers were involved in the program. By the end of 1953 more than three hundred million parasites had been released into infested Mexican citrus groves.

The Mexican government was impressed by both the success and the economy of the citrus blackfly project. In January 1954 they set up a biological control department within the Defensa Agricola. For the first few years of this department Smith was more or less in charge, in fact if not in name, of planning, organizing, evaluating and conducting its work.

In February 1954 Smith wrote congratulations to his daughter in Paris on the birth of her third child. He wrote that he was busier than ever on the blackfly work. Herb had been loaned a new De Soto passenger car by the Mexican government and was able to cover much more ground than previously. His previous vehicle had frequent breakdowns, often on remote country roads far from help. He was driving about 8,000 miles a month most of the time.

Smith was going to reach his 60th birthday on June 29th and wondered how he would feel. He wrote that "I don't think I'll even notice it. I am keeping up with men much younger than I am, 25 to 30 years old, and they seem to like being with me. At least they are all after me to go with them places."

In another letter home, Smith reviewed all of the troubles he was having with the various Americans he had to deal with. He felt none of them were supporting his work, and were instead making the job more difficult.

> *Well, things are really coming to a head. Dr. Porter was here the week before last. I am only thinking*

of controlling this blackfly down here, no matter who I have to lick, meaning Baker and his crowd. Very few men would ever tackle Baker. The only one who did was a former Chief of Bureau and he could not fire him, so he sent Baker to Mexico when they were killing lots of people and he thought he would quit. Baker didn't and he is still here.

Baker is doing everything he can to interfere with this parasite work and slow it up. If he had the parasite end of it, as he tried to get (the Chief gave it to Clausen) he would nurse it with even a bigger force for the same length of time he has nursed the fruit fly here – 23 years and they are still piddling with it – every year the same old experiments with a slightly different twist.

The Mexican officials even think it is a joke the way they have hung on and still no results. The fruit fly is worse every year. You can see how they all feel, especially Baker. I come down here and in three years control my pest – one American and flocks of Mexicans who work for the Mexican government (not our government as Baker's Mexicans do).

I take what the government gives me and get things going. The Mexican government thinks the world of me, the only entomologist that ever really worked with them. They wrote Hoyt just this. I hear this from all sides. Baker looked down his nose at them. I jumped right in with them overlooking their inexperience!

I have not been to lab but once two weeks ago, and did not go to my room. I knew there would be a note from Baker. My other mail there is only advertisements so I don't care. Yesterday I just thought I would sneak up the back way and look through windows. On my desk was a copy of a letter to Hambleton and the subject was – Herbert D. Smith.

Naturally I did not go in as I would be mad as hell as he probably threw the book at me. He may have dug up every little infraction of regulations against me and more. The Mexicans hate his guts and are behind me. I told them that I may have to resign and will work for them till the job is finished. They said "Don't worry, you won't have to retire." They will get Baker out of Mexico first. They are only waiting to see how far Baker goes.

If I have to retire it will be, in a way, a small international incident. Especially when I quit because our government wants to spray and the Mexicans will have no part of it. If I work for the Mexican government it is probably the end of the lab here. It may be the end of it anyway. That is why they are after me in every way at the lab.

All of the Mexican government is now on their annual vacation so I did not want to read Baker's letter. I would be mad all week and write something to Washington that would do me harm. I will leave the letter on my desk until June first when they return, and then take it down to the officials and their Secre-

tary of Agriculture or even higher, which ever they want to do.

Of course the letter may only be that Smith doesn't come around to the lab much anymore and we would all like to hear what he has to say on the work and so forth. But if Baker saw the letter sent to the Chief by the two Agricultural Agencies here, he is going to be boiling mad and shoot the works. He always does.

Smith had been given a copy of the letter the Mexican blackfly committee had sent to Washington praising his work. The Mexican Bureau of Entomology had an even more strongly worded one. Three weeks after the Mexicans contacted Washington supporting Smith, Hambleton told Herb that the Chief hadn't received any letter from the Committee, as the Chief always circulates letters to the divisions when it concerns them.

Hoyt was sitting on the letters. Smith felt Hoyt "believes in Dr. Baker too and is probably wondering which way to jump." Smith hadn't yet seen the letter from the Mexican Bureau but would take a look when they returned. While his spoken Spanish was good, Smith had a local medical student who had gone to school in Jersey City, who translated and typed official correspondence.

Even if I get a letter from Washington this week and it isn't what I like I will do nothing till the Mexican officials return. I can easily be on a trip, No telephone calls are to be given me except from certain peo-

ple. These are three men of the Committee. This is to ward off getting someone from the lab on the other end of line. I might tell them off, and I don't want to give them that chance. Anyway I have moved around a lot and think I have thrown them off the trail. When the telephone rings I usually never answer it, or go down to office and ask them to ask who is calling.

To think that a government man has to come to all this, like something out of a detective novel. I really do not know why they should hate me so much at the lab. Just because I don't have anything to do with them shouldn't make them that way. You know me for not wanting to bother with people who are boring etc. But they do hate me and go out of their way to hurt me. Anyway this week or next should tell what is going to happen here. I think I am ready for it.

In January 1954 the Mexican fruit fly was discovered south of the US border near Tijuana. This set off alarm bells among American agricultural interests. One consequence of this was that more money was alloted for the fruit fly project at Baker's lab in Mexico City. Their interest in getting control of the blackfly somewhat lessened at the same time.

On May 12th, Herb wrote to his family that he had just returned from a trip around the west coast with Curtis Clausen, now with the University of California at Riverside, Ted Gardner and his new boss Paul Oman.

Things worked out very well here. My work won all around. Baker got set back everywhere on the blackfly. Cooper is soon finishing on the west coast. Both Mexican departments backed me to the hilt. All the talk at the conferences was about me and my work. The new chief was also convinced on the parasite work and spending less money and letting the Mexicans do it.

Fortunately for Baker, the fruit fly that he has been working on for past 24 years here was recently found at border of California on the Mexican side. The blackfly now takes second place by California and they are pushing fruit fly and allotting $50,000 for the work. Baker comes in on this and almost overnight his men left the blackfly work and got onto the fruit fly. It shows that the blackfly work was just to keep them on the payroll since they were able to stop the work so suddenly.

No mention was made at the meetings about Smith avoiding dealings with the men at the American lab. It was Baker who was criticized for not cooperating better with Smith. Everyone knew Herb had a resignation letter all made out and was prepared to quit if push came to shove.

Stone and Shaw tried to run down my work at a small conference in the hotel room, but Clausen backed up the work and this shut them down. I think they were all prepared to ride me and my work as be-

fore, but they soon saw how the wind was blowing and also the men, Clausen, Gardner and Oman, had just seen the results of parasite work from Texas on down to Mexico City. They had the same opinion as I have and were even more convinced.

Washington apparently does not want me to quit. They let me publish my work separately, not with Baker, and I never even asked for that. Baker's work will only be spraying which was of no use and a failure here, so he gets nowhere. My work has cleaned up the pest.

Gardner kept fishing around to find out how I felt about the future and he finally said "Well, you just do not care whether you work for the government or not." I replied right back that was just it. So they all know how I feel. Things have changed a lot in the parasite line – almost wiped out. I am supposed to help them in their new Biological Control Department.

I imagine the men at lab hate my guts. I wrote a stiff letter to Washington saying that Baker and his crowd were nursing the work along and trying to make it last. I also wrote a lot of other stuff as I did not care whether I quit or not. It went to the Chief too. I also said I would meet that crowd anywhere in Mexico and settle the matter but with Mexican government officials in my corner. Not one word was said about the letter. I think it opened their eyes in Washington.

So now that the bosses are gone, I wonder what I am in for at the lab. Not that I care as I do not even have to go there. If Baker starts anything I will warn

Washington that I am through unless they do something. Baker must know this.

On December 16th Smith wrote home from Valles to apologize for not being able to get home for Christmas. The parasites were doing very well and many fields were clear of the pests.

The best ones from India are now hitting their stride. Zaralla says everyone now is for parasites. Right now collecting is poor because it is winter. They now have three times as many collectors as I had. Even Petersons are all for them despite Cooper who says they aren't worth anything. They'll soon see and the lab will be closed!

I have a lot of traveling to do. I haven't been on West Coast in months. The National Committee pays no attention to Baker anymore – it is all me because of the success with the parasites. I bet he wished I was still attached to his lab as he could grab off the credit.

In May of 1955 Herb received the surprising news that the U.S. Department of Agriculture had decided to honor him for his work in Mexico. Kipling sent Smith a letter stating "I am extremely glad to advise you that the Department of Agriculture has recognized your outstanding contributions in entomology research by awarding you a medal for superior service. The work you did in the introduction and establishment of parasites for the control of the citrus blackfly is one of the most outstanding examples of the importance of natural enemies in the control of destructive insect pests."

Herb wasn't sure how to react to this development. He wasn't too pleased with the people who he had to deal with in the department, and was more used to butting heads rather than schmoozing at some event. His first thought was to decline, but he was practically ordered to go.

His boss Dr. Oman who had been running down his work in Mexico must have found it awkward to have had to write a follow-up memo to Herb stating "I wish to congratulate you for the fine work that led to your receiving this award. In the event that you cannot attend, please let us know, but remember that we are looking forward to seeing you here."

Herb wrote about the situation to his daughter Janet in Paris.

> *I bet you could have knocked both Knipling and Oman over with a feather. They gave me a bad going over, especially Oman, last winter and I was ready to tell them where they could stick their job. I didn't as the Mexicans could not finish it without me. Now they can. I was about prepared to resign June 30 because of Washington's attitude.*
>
> *Gardner has been constantly advising me to get out while I had good offers, one in Ceylon for UNO (FAO)* which would bring me much more salary*

* Food and Agriculture Organization, UN agency dedicated to eliminating hunger and improving agriculture.

as I pay no income tax on it. They were after me as Clausen and Harry Parker both recommended me and maybe others.

However, when they asked the Entomology Research Branch (previously the Bureau of Entomology and Plant Quarantine) Knipling wrote them a long letter saying he could not release me as I was too important – and that after the treatment I got. Knipling is a spray man and Oman is a dead-insect man and apple polisher. He would not back me up on my work down here probably because it was done under OKs of three other bosses – Clausen, Hambleton, and somewhat under Gardner. They okayed my plans as they went along but Oman said I did everything wrong.

In other words I should have knuckled under to Baker and his spray crowd. I fought them and chased them all over Mexico and closed their work out – with the Mexican government's assistance. Oman thought I should have discussed every move with him and his men! What a joke. They were trying to push me off the job and ran my work down to every one.

Baker and his men had to stop all blackfly work last November. Boy am I popular there! I told Washington if they kept digging up old scores I'd call for a showdown, with the Mexican government officials in my corner. Now since my return from Washington they don't mention the past and what I did to them. It hurts them even to talk to me. I have my office with the Mexican government so it does not bother me. They held up my raise in grade for two years because of this. I am now level 12.

Why I wrote is mainly to say I have been selected to receive a citation of "Superior Service Award." Parker will know about it. It puts me in very distinctive company. Oman, Gardner, and Clausen never got it. The Vice President will hand out the citation and medals. About 15 or 30 men out of 100,000 in the whole Department of Agriculture get it.

I tried to duck out. I thought it was just another copper affair they pass out to everyone, like attending Sunday school. But they put me right rapidly. I got an LA through the fastest ever. They wouldn't pay me $12 to attend a conference last winter from home. I paid my own way. Now, $300 to $400 just to pick up a medal! They even wired it down.*

I leave Sunday by plane for Washington and home, then drive the family down for the ceremonies on Tuesday at the open air theater near the Washington Monument. They seem more enthused about it than I am. I was hoping I didn't have to go and that the LA would be delayed like all their other papers. I still have lots of work here and I don't like to leave now.

Maybe I am the only one from the Entomology Research Branch. If I am that will explain why they want me. They sent out engraved invitations, five for

---

* A governmental travel voucher.

the immediate family and close friends. I sent two to Lee and he may be there.

It came from high up down. Probably Secretary Benson as he was here and talked with the Mexican Secretary of Agriculture. He undoubtedly mentioned my work as it is in all the papers, off and on. Also I run their National Committee of 1,500 people. They spent US $1,000,000 last year and the year before and that ain't peanuts. 70% of it is my work. Blackfly is successful everywhere now in Mexico. Only one State is not entirely controlled, due to lack of parasites and attention.

FOA offered me a job in Bolivia but I refused – too rough and too high. When this citation gets out I may get some offers. I told FAO (get the difference?) that they not worry about what the Bureau of Entomology writes, that I was my own boss and prepared to resign and accept. Then they asked to check me for loyalty (strict for UNO) and said it would take three to four months. I said okay but do it while I am in my job. They wanted me to resign first. Then they wrote that Ceylon withdrew their request!! What administrative work! Lousy! Suppose I had resigned – out on a limb – what? I think they were afraid to take me away from the Bureau of Entomology as they borrow lots of their men. They borrowed Berry. The Bureau of Entomology released him easily.

Maybe they will treat me better this time. If they don't I am going to tell them what I think and resign. They don't like to control things – pests – as I have done. They want experiments, nurse it along, make it last, write up bulletins. None of these do I like and that is why I succeeded.

Mexico wanted control and fast. They paid for it. I suppose Parker will be sorry if I don't write it up, my work here. But he won't be the only one. I like to do the work and get on to the next job. Let the stenographers write it up, from my reports.

I may take a few weeks vacation and see Lu graduate. I never saw you or Pat graduate high school.

On June first at the open air theater next to Washington Monument, the United States Army Band played the Star Spangled Banner, the Senate Chaplin Dr. Harris gave an invocation, Congressman Clifford Hope gave a speech, and then the Secretary of Agriculture Ezra Taft Benson presented Smith and the other honorees with their awards. The award given to Smith read as follows:

Superior Service Award to
Herbert D. Smith
for exceptional diligence and perception in the use of parasites to control the citrus blackfly in Mexico, thus contributing to the protection of the citrus industry of the United States.

The presence of the citrus blackfly (Aleurocanthus woglumi Ashby, family Aleyrodidae) in Mexico since 1935 and its continued northward spread represented a serious threat to the citrus industry of the

United States. It was deemed advisable to try, by all possible means, to prevent the entrance of this pest into the United States. Because the economy of the citrus industry in Mexico could not support extensive and continuous control of the pest by chemical means, the use of biologic agents to hold infestations at a low level and lessen the possibility of spread assumed great importance.

In 1949, Herbert D, Smith was sent to southeast Asia, the native home of the citrus blackfly in search of its natural enemies. He found a number of potential enemies in India and Pakistan and introduced them into Mexico. The following year Smith was given the responsibility of studying these parasites in their new environment, collecting the progeny of established species, colonizing them in widely separated citrus growing regions of Mexico where the blackfly had become established, and observing their effectiveness in controlling the pest under diverse ecologic and climatic conditions.

Herbert Smith approached the monumental task before him almost single-handed. The Mexican agency charged with the control of the citrus blackfly had no personnel trained in the biological control of insects. Thus it became necessary for Mr. Smith to convince the proper authorities of the value of the proposed program, and to train Mexican personnel in methods of collecting the parasites in the field at the right time, packing them for shipment to other regions, releasing them in areas infested with the citrus blackfly and lat-

Herb Smith receiving the Superior Service Award from the Secretary of Agriculture, Ezra Taft Benton, in a ceremony on the Washington Mall, June 1, 1955. The award was given to only a small percentage of Department of Agriculture employees. It was a surprising turn of events for an agency that Smith felt was rarely on his side in the fight against the citrus blackfly in Mexico.

NATION'S GRATITUDE—Secretary of Agriculture Ezra Taft Benson, right, presents the department's Distinguished Service Award to Herbert D. Smith, 5 Carmita Avenue, Rutherford, in a Washington ceremony, yesterday. Smith was honored for "diligence and preception in the use of parasites to control the citrus black fly in Mexico, thus contributing to the improvement of the citrus industry in the United States".

er observing these areas of colonization for extent and degree of parasitization in order to determine the effectiveness of the parasites. In order to efficiently train personnel, he learned to speak the local language, thus further gaining the confidence and respect of Mexican officials and workers.

That Mr. Smith was eminently successful in attuning his objectives is amply attested by the fact that he quickly gained the complete confidence and cooperation of the Mexican National Committee for the Combat and Control of the Citrus Blackfly.

Further, the overall results of the use of parasites for the control of the citrus blackly have been so striking that the Mexican Department of Agriculture has created an organization to investigate the possibilities of controlling other insect pests in Mexico through the use of parasites, and has formally requested that Mr. Smith be authorized to advise and collaborate with that agency.

In the biological control of the citrus blackfly, Mr. Smith accepted the problems as personal challenges, and the manner in which he met them is representative of the highest type of scientific initiative and devotion to duty. He ignored personal convenience and comfort and frequently worked "around the clock" in his efforts to supervise the program or attend pressing operational matters in various parts of Mexico. Due largely to his personnel efforts, suppression of the citrus blackfly by biologic means throughout the great-

er part of Mexico is an accomplished fact months or years in advance of the normally expected time.

For this accomplishment, far beyond the normal responsibilities of an assignment, Herbert D. Smith is deserving of the highest award from American agriculture.

The wording of the award was a surprisingly frank and true account of the blackfly situation. Given all of the conflicts Smith had been having for years with his American colleagues and supervisors, one wonders who wrote it.

Herb suspected that Lloyd Benson, the Congressman from the Rio Grande Valley, was behind it all. Benson had been down to Mexico meeting with the leaders of the Mexican Agriculture Department. He thought the Mexicans pressured Benson to make sure the Americans supported Smith's work.

After the award ceremony Jean decided to try life in Mexico again. She spent the summer at Herb's apartment in Mexico City, but his work continued to require long absences from town, traveling around the countryside monitoring the citrus blackfly. That left Jean a choice between staying alone in the city or traveling with Herb on dusty road trips. Jean found that neither dust nor the altitude of the city were good for her health. She decided to return to their home in New Jersey and wait for Herb to finish his work in Mexico.

In September of 1955 the East Coast of Mexico was ravaged by three separate hurricanes arriving in quick succession. The first, Hurricane Gladys, was weakening when it made landfall on September 6th but still caused torrential rains and flooding all the way across to Mexico City. Hurricane Gladys hit the coast on the 19th with 160 mile per hour winds and adding to the flooding that had not yet cleared from the prior storm. As if that wasn't enough misery, Hurricane Janet arrived at Veracruz on the 29th after earlier sweeping across the Yucatan Peninsula.

The hurricanes caused hundreds of deaths in Mexico and displaced many thousands of people. The United States in response put together a massive relief effort to aid Mexico. While of lesser importance than the direct human toll, the storms did also put the blackfly parasites to the extreme test. For months afterward the Valles-Tampico area experienced a severe fluctuation in infestation and parasitization.

In December 1955, Herb wrote in a Christmas and New Year's greeting to his daughter Janet that he again had to spend Christmas away from home. He was considering to at least try to get over to Acapulco for the holidays. He said the parasite work was going so well that he was able to spend some time on other parasites for scale insects. "Mexico is going to build a big biological control insectary and laboratory this coming year. They are all sold on parasites."

Herb continued extensive travels around Mexico throughout 1956 in cooperation with the Mexican National Blackfly Committee, Control in the areas south of El Carmen, near Ciudad Victoria, in the Northeast and

Guaymas in the Northwest, was by parasites only – spraying was permitted only with the Committee's authorization. The areas northward from there were considered the chemical control or eradication zone.

Biological control improved in every region where it was protected from spraying. More than 25,000,000 parasites were collected and liberated in other regions. By the end of 1956 the hurricane-caused disturbances in blackfly infestations in the Valles-Tampico area had resolved and the blackfly was again under complete biological control,

Smaller fluctuations were seen around the rest of Mexico, but these were considered normal and of no commercial importance. The arid Northwest continued to be the most difficult area to control in all Mexico. Only 10% of the groves had infestation enough to warrant further parasite releases. Experiments with parasites in the eradication zone near Guaymas gave good results showing that the parasites could work there as well as elsewhere.

Smith was also helping the Mexican biological control department to move beyond just blackfly and to try out biological controls for other insect pests within Mexico. One of their projects, importing a variety of parasites from Hawaii to attack the Mexican fruit fly, put Smith back in conflict with the American lab. The results however were less impressive than achieved with the blackfly, partly because there were not enough funds and personnel provided for a large effort.

A number of other species were collected and released against other insect pests. There was *Casca smithi*, a parasite of red scale obtained from Hong Kong, *Che-*

lonus heliopae and *C. narayani,* parasites of the pink bollworm, were reared and released into cotton fields around Torreon. The native parasite *Aphelinus mali* was collected around Mexico and distributed into new regions. Purple scale was being well controlled in release areas by the parasite *Aphytus leptidosaphes* obtained from Texas.

Smith and the Mexican biological control department also offered assistance to other countries. Ecuador was one – they had been releasing *Amitus* parasites sent from Mexico onto citrus blackfly populations there, and were sending samples back for analysis in the Mexican lab.

A spittle bug of pasture grasses in eastern Mexico, *Aenolamla gostica,* had resisted all attempts at chemical control over a three year effort. Local dairymen asked that biological control be tried against the pest known locally as *mosca pinta.* A reduviid predator native to Mexico had generally controlled the pest in the past, but nearby cotton dusting and the chemical treatment of the pastures appeared to have upset the natural balance. The hurricane and flooding of 1955 worsened the situation. The damaging nymphs feed very close to the soil and cannot be reached by insecticidal dusts, and the spittle offered some protection as well. Entire fields of pasture are destroyed by this pest.

The Mexican government set up a new laboratory in Tampico for the biological control of the spittle bug. The reduviid was being reared there in fair numbers for release in the coming spring when the first nymphs appear.

The Defensa Agricola also authorized the expenditure of US $20,000 to build a Biological Control Laboratory on the outskirts of Mexico City, with construction to begin in the spring of 1957. It was to be of an advanced modern design and would handle nematode work as well as the insect biological control activities.

In 1956 the Mexican National Blackfly Committee continued much the same as in 1955, spending US $851,289. Of this a majority was spent on the biological control program.

Smith was asked to continue to consult on citrus pests in the Rio Grande Valley of Texas. In 1956 he made five trips there to study the parasites in the Valley citrus groves. Though he was effectively banned from introducing parasites there, he could study them. Smith worked closely with Herbert Dean of the Texas Agricultural Experiment Station in Weslaco. Dean was in charge of biological control studies for the state.

About twenty blackfly infestations were reported in the Rio Grande Valley of Texas that year. Around half of the material showed considerable parasitization by *Amitus* with rates as high as 90%. The parasites must have arrived together with the blackfly, possibly on infested leaves or blown across the border from some hidden source. Smith had noted in Mexico that both the blackfly and its parasites could travel at least 40 miles in a strong wind.

There was a general feeling among the growers and others in the Valley that chemical control for eradication might not now be necessary because of the effective work by parasites. Nevertheless, Smith and Mr. Dean, the biological control specialist for Texas, advised that the present spraying program be continued for some time yet.

It was also decided that if further infestations were found, Smith would be notified and he would immediately go to Texas to study conditions in the field and rear out material under quarantine conditions. The door was left open for future releases of parasites from Mexico.

Unlike in Mexico where Smith worked hard to get everyone to cooperate in keeping insecticide spraying out of the biological control areas, in Texas parasites had to fend for themselves when nearby areas were sprayed. It was obvious that this created problems. Smith observed that most scales were well controlled by their parasites and predators where there was no inter-planting of cotton with consequent insecticide dusting.

Smith noticed the Florida red scale increased in great numbers during the summer but was under control by its most important enemy, the twice-stabbed lady beetle, by the end of the year. He hypothesized the buildup of this scale may have occurred when the lady beetles moved to cotton in the late spring to feed on aphids and other insects. Insecticides applied by crop-dusting aircraft to combat these aphids killed many of these beetles as well. The scale parasite *Aphytus lepidosaphes* continued to spread around the Valley. The rust mite, which had few natural enemies remained the main problem.

Other organizations doing international pest control projects expressed interest in hiring away Smith from Mexico. In January a group came out from Washington to try to talk Smith into staying on the job in Mexico. Point 4 was one of the groups that discussed Smith's availability with the Department of Agriculture. Point 4 was a technical assistance program for developing countries created by President Truman. It was intended to spread economic growth and compete with communism for the "hearts and minds" of people in the third world.

Point 4 had been having discussions with Herb for some time. They were interested in having him spend a few years in the Philippines or Thailand, setting up biological control programs there. Smith put them off as the Mexican project still needed him. But Point 4 continued to pursue the matter.

> *They saw the Agricultural Attaché, thinking he would take them to the Point 4 man. He said definitely not, that his department could handle it and keep me at same place. It's the same friction there between the trio. So now they will have another conference in Washington to see about who I work for. They talked with Oman twice by phone. I would have made out much better in Point 4. Mexico had their Ambassador write to the State Department and see them, and the latter wrote Benson about me. So they got it from all angles. I only wish they had left me alone so I could retire.*

For all of the conflicts Smith had with the Department of Agriculture about his refusal to play ball with their bureacratic ways, the agency knew how valuable he was and tried to keep him on the job.

Strange how everyone is after me to stay here – Stone, the Pest Control people, and Defense Agricultural Committee, all spoke strongly about me. I was at the conference conclave too when they talked about me. I guess I have done a good job here but the push seems gone out of me.

In April 1957 Smith wrote home that the Mexican government was stopping much of Stone's spray work in order to protect the parasites. "It's no longer a fight now. I am out of it. What the Mexicans do is not my fault anymore. Stone and the rest of them know I will quit at the drop of the hat, and they don't want me to as things would be much worse."

The citrus blackfly was continuing to be well controlled in all the areas of Mexico under biological control. In one three month period, April to June 1957, about 11,000,000 adult parasites of *Amitus hesperidum, P. clypealis* and *P. opulenta* were collected for liberation.

Only sporadic minor infestations were found in a few areas. Each year the situation was becoming more stable. *Prospaltella clypealis* remained the primary controlling factor in most of Mexico. In the northwest, however, *Prospaltella opulenta* was the only effective parasite with *Amitus hesperidum* a poor second. *P. opulenta* was also increasing greatly in the Victoria area and Smith felt that it might be on its way to becoming dominant there as well. The city of Guaymas was technically clean of the blackfly due to *P. opulenta.*

In July 1957 Mexico City had its worst earthquake in 48 years, killing scores of people and injuring hundreds. Until then Smith had been lucky in avoiding such troubles in the temblor-prone region. Colima and Jalisco both had severe earthquakes in 1932, and Colima had another in 1941.

Defensa Agricola had a new Director named Ing. Benjamin Cortina Carmona. Fortunately he was very interested in biological control and was pushing for its use in Mexico even more than any previous Director. In September Smith and Ing. Jimenez took the new director on a tour of the Tampico area to show him the parasite work with the blackfly as well as the spittle bug and other biological control efforts.

The year-end report from the Department of Agriculture on the Line Item "Biological Control of the Citrus Blackfly" indicated that the expenditure was reduced in the last half of 1957 and terminated at the end of the year. The report stated that "no Station Annual Report has been received from Mr. Smith, and in view of present plans for preparation of a report on the project, no special request has been made. In lieu thereof we are transmitting a copy of Mr. Smith's report for the second quarter, received in August of this year, supplemented by excerpts from a memorandum dated September 15, 1957. It is believed that Mr. Smith's activities since these two items were prepared have been concerned primarily with work for the Plant Pest Control Division to which he was detailed effective July 1, 1957."

In early 1958 Smith published an updated report on the blackfly work. Again he choose to publish in the Spanish language in the Mexican journal Fitofilo. On one level this was the logical choice. It was the Mexicans with Smith working along side who had done the work. The Mexicans rather than the Americans were most interested in the work. But one might assume he wouldn't be bothered if some of his American colleagues took it as a snub – a well-earned one.

In March of 1958 Herb received a letter from Oman saying he would be placed under the control of Stone's fruit fly lab. The wording of the proposed assignment made it quite clear that the other entomologists in the agency were still bothered by Smith's independent approach in dealing with the Mexicans in the biological control field.

Oman's letter said that "Administrative and technical supervision of Mr. Smith's activities will be with the Entomology Research Division. General plans for activities for Plant Pest Control to be undertaken by Smith will be cleared through the Entomology Research Division. Details will be cleared with W. E. Stone of Mexico City Lab. Stone will have immediate supervision over Smith's activities. Stone should also be notified of any suggestions made by personal contact between representatives of the Plant Pest Control Division and Mr. Smith in the field." To Smith it was another clear attempt to restrict his freedom of action and ability to get things done in Mexico.

Oman doesn't like me because I am no "yes" man. He quotes a regulation that my work does fall into Stone's division since I don't introduce parasites into the US. It is the old government crazy stuff that each Division leader wanted his own parasite work. They are all insecticide people and don't want biological control to do away with their work. They want it to last as only insecticides can do. It's the old government game. Nurse the job along and keep your nose clean. I think Stone would have preferred I stay as I was. He seemed to treat me pretty decently the past year or so.

The difference between what I earn now and if I quit is very small (with my government pension). In other words I would be working and fighting under Stone for very little extra money. When I quit I can easily earn a good salary and the Mexicans will pay me well if I went with them.

Smith wrote home that the only choice was to resign. He had already told Oman of his intentions and it would only require an official notice. If they wouldn't let him work as he had been doing so successfully for the past seven years, what would be the point of staying on? Smith expected to leave Mexico for good by boat in June. He hadn't told the Mexicans of his decision yet, and he wasn't sure what they might offer to try to keep him in Mexico.

They have to put me in Baker's lab, actually Stone's lab, as Baker retired a year ago. Several of those men still hate my guts. The others probably do but don't say anything. I refused to accept the position. They cooked it up anyway. They want me to stay here awhile and help them as I have such good contacts in the Mexican Government and they don't.

If I took this job I would no longer be in the Parasite Division. Washington asked me to reconsider. I would be under Stone and must do what he tells me to do. He only nurses the work along, doesn't want results or control as I do. He has been working on the same insect – the Mexican Fruit Fly – for 27 years and the fly is worse than ever.

Another division wants me to stay here too to help them. Also the Mexican Government wrote letters to their Ambassador, to Benson, and other big shots to keep me here. Everything was fixed for me for two more years here, when they finally sent the details of the assignment which put me under Stone. No word was ever mentioned about this before.

So I have sent in my resignation. If I had a pleasant job as I now have, being my own boss, reporting directly to Washington, I would stay. I don't like that man as my boss, and he hates me. Clausen resigned for a similar reason several years ago. Washington isn't interested in biological control, only insecticides as it is a billion dollar business. All parasite men are stuck under insecticide men so they won't wipe out the work as I did here.

The Mexican Government may do something. They want me and need me. They pay low expenses etc. but we'll see. If I worked for the Mexican Government I'd sure make the Americans here hop. Or maybe I'll write about government service for the Saturday Evening Post. They hate government service and say it is lousy, which it is. '

Anyway I'll be 64 this June. Why work all my life? No more headaches, and high blood pressure, fights with Baker and the whole lab. They still hate me and one man always greets me in front of everyone with "there's the man who put us out of our jobs." I didn't, the Mexican government did. But I helped as they were just nursing it along to make it last.

This fellow had a swell villa on the beach here at Guaymas, rent paid. It's no wonder he hates me. But I pay taxes and don't like loafers. After I get out I will contact the Saturday Evening Post who think badly of all government work in general and may write an article.

Biological Control is at its lowest in years. Everyone is afraid to put up a fight for it. It is more important here than it is in Washington. Not on your life do I go there and work under Stone. Well, I'm glad it's all over but the shouting. I should have quit three years ago and gone with Point 4, but stayed to help the Mexicans.

Smith ultimately decided to deal with the bureaucratic reshuffle and stick it out a while longer. He was still training some men to take over the work and didn't want to leave that unfinished. He spent the month of July on the West Coast of Mexico showing them the ropes. He was suprised to find that "the new work is just the same as before as far as I am concerned. The fact is that Washington bothers me lots less. My new office doesn't bother me at all. So in a way it's even better, but I still think I will retire in January."

Smith was starting to make plans for his retirement, He had seen the massive blackfly control effort through and it was a remarkable success. Herb was 64 years old and had been battling the elements and bureaucracies and those who opposed the work for a decade and a half. The Mexicans were well able to take over the work now. He was thinking about buying an orchard in California or Florida. He could relax a bit and finally get some time to spend with his family.

In January 1959 Herb was not feeling well, He had gone to a doctor who gave him a liver function test and told him his liver was "slightly out of kilter" and that he would need to take some sort of medicine. Whatever he was given may have made him worse. "This toxic stuff gave me an infection in my chest and made my heart go awfully fast when I moved around. If I walked upstairs slow I was puffing so bad I had to lie down." The doctor had him come into the medical office every other day to get "some medicine directly into my veins to clean out my blood of toxins. I give him some whiskey."

Eleazar Jimenez Jimenez was an entomologist who often worked with Herbert Smith on the citrus blackfly control project. He was one of the few Mexican scientists with some training in biological control, having participated in a course at the University of California. Jimenez went on to become the head of the government biological control programs.

The citrus blackfly campaign would be officially disbanded at the end of the month. Whatever work that remained to keep the pest under control would be up to the Defensa Agrícola. Herb thought it was "rather sad – all the people – lots will be laid off, and the men in the field. I had expected this for some time but not so soon." The work would be supervised by his old friend Eleazar Jimenez Jimenez. Herb had known the new program director as well for years. He felt it would be in good hands.

At around 10 AM the morning of Tuesday, January 20, 1959, two Mexican patrolmen driving in a radio car along the park known as the Alameda Central reported seeing a man lying motionless on the grass. The man was Herb Smith. The Alameda park was next to the Mariscala building housing his Mexican lab. The officers brought

Herb to their car and drove him to Hospital Juarez, but doctors there were unable to revive him. The newspapers stateside reported Herb had died of heart attack, but physician Dr. Miguel Cervantes certified the cause of death was cranial trauma.

Herb's family was alerted by phone of the tragic event. They were looking forward to his retirement and return home and were devastated as well as confused about what had happened. His wallet was not missing, suggesting he was not the victim of a mugging.

Given the difficult relationship Smith had with the forces opposed to his biological control work, there was speculation whether this could have caused Smith to have become the victim of foul play. If so, the timing seems illogical, as by the time of his death the biological control effort of the citrus blackfly in Mexico had won and was beginning to wind down. Herb had largely handed off control of the project to his Mexican colleagues and was instead looking forward to retiring and returning home.

It is possible Smith collapsed due to a medical emergency and hit his head in the process, but as the officers reported finding Smith on the grass rather than on pavement it is unclear how the fall could produce such cranial trauma. If Herb was a crime victim it would be a surprising and ironic turn of events, as he had spent over a decade traveling hundreds of thousands of miles back and forth all areas of Mexico without incident. What actually happened may never be known.

Herb sadly never made it back home to his family in Rutherford and retirement. Mexican colleagues organized a funeral for him attended by the many friends he had made in Mexico. The American vice consul then arranged for his final return to Rutherford. On August 17, 1959 the Mexican Entomological Society and the Department of Parasitology of the National School of Agriculture held a "Dia del Insecto" symposium at Chapingo in Smith's honor. There are reports that the Mexican government placed a bust of Herb in a park commemorating his work there.

11

Epilogue

Herb Smith spent more than a dozen years of his life on the citrus blackfly problem in Mexico. The control effort had succeeded by the time of his untimely passing. In the ensuing years the parasites that he introduced into Mexico have largely controlled the pest there. As expected in dealing with complex interactions between various species and ecosystems, there have been occasional setbacks and local flare-ups of the citrus blackfly, but these have been manageable. Weather events such as freezes and hurricanes in Mexico have caused disturbances requiring augmentation of parasites controlling the citrus blackfly there.

In October 1959 one of strongest hurricanes ever recorded arising along the Pacific Coast devastated western Mexico. Worst hit were the citrus-growing states of Colima and Jalisco. At least a thousand people died in the storm and another 800 died in a mudslide caused by the heavy rains. The high winds caused a loss of parasites and a resurgence of the blackfly. In the aftermath of the hurricane the Mexican government agronomists who took over the control effort were able to reintroduce more parasites and reestablish the equilibrium between pest and parasite.

The Mexican government has continued over the years to maintain control of the blackfly and to expand its efforts to control other agricultural pests using natural enemies. One of the key Mexican scientists trained by Smith in the biological control of insects, Eleazar Jimenez Jimenez, headed the Mexican biocontrol efforts and has been termed the "father of Mexican biological control." Many other scientists have continued the work there. These entomologists include Hugo C. Arredondo-Bernal and Luis A. Rodríguez del Bosque. The Sociedad Mexicana de Control Biológico was formed to encourage scientists working in the field there.

This interest in biological control has continued not only in Mexico but in many other countries, especially in Latin America and Asia. Many universities and government agriculture departments have sponsored much research in this area. The field unfortunately has not fared as well in the United States. The large Department of Biological Control that the University of California operated from its Riverside and later Berkeley campuses began to lose its aging scientists and was eventually disbanded. The biological control library largely compiled by Curtis Clausen was discarded in the process.

In 1991 the European Parasite Laboratory was combined with Rome-based Biological Control of Weeds Laboratory, established in Rome in 1958. A new laboratory was built near Montpellier, France, with satellite laboratories Rome and Thessaloniki, Greece. The USDA laboratory in Hoboken was moved to Port Elizabeth in the 1990s and the old building demolished to make way for a new office tower and hotel.

The home base for Herb and the Mexican biological control staff, the 20-story Mariscala building near the Alameda, in Mexico City was heavily damaged in the devastating 1985 earthquake. It was condemned and later demolished.

One of the parasites that Herb studied in Europe, *Dibrachoides dynastes,* was collected and released against the alfalfa weevil in the 1980s. The parasite did not do well against the winter weather of the eastern U. S., but has become established in California and Wyoming. It has not been as effective controlling pest populations though as have the blackfly parasites. Scientists are still looking for better alfalfa weevil controls.

The fight against the European corn borer continues, with biological agents keeping the pest partly in check. The parasite that is widespread in Europe and that Smith studied most intensively, *Phaeogenes nigridens,* does not seem to have received much attention in the U. S. in the years hence, but a number of other parasites have been used against the pest.

Mexican agriculture has continued to expand and the United States is a major importer of these products. Economic development however has not come to all the Mexican people. The backwater town of Badiraguato where Smith started the blackfly control effort in 1943 has struggled economically, especially along the eastern hills where Smith found the citrus groves. The mayor was quoted in 2010 as saying "up here in the mountains, there are no schools, no roads, no electricity and no water other than that provided by Mother Nature."

While Herb hoped that a thriving citrus industry could bring prosperity to the Mexican people he worked so closely with for many years, other products proved to be more lucrative in the American market. Until his arrest in 2014, Badiraguato-born Joaquín Guzmán Loera, head of the Sinaloa Cartel, was considered to be the most powerful drug trafficker in the world. The nearby city of Culiacan, where Smith spent much time colonizing blackfly parasites, has long been the headquarters of the Sinaloa Cartel.

The citrus blackfly has generally been controlled in most areas of the world and usually causes little economic damage. The main controlling agents are parasites. There have been several episodes where the blackfly has made its way into the United States despite all efforts at quarantine. The blackfly has made it in the Rio Grande Valley of Texas on a number of occasions. These outbreaks have all been rapidly brought under control.

An outbreak of blackfly occurred in the Fort Lauderdale area of Florida in 1976. The state released parasites in 700 locations, but at the same time pushed for local application of insecticides. The director of pest control for the state, Charles Pocher, was quoted as saying "parasites will not eradicate; they will only reduce populations and give us a little better chance to eradicate."

While it is true that parasites control rather than eradicate a pest, that is a rather narrow view of the situation. As Herb found out in difficult months of searching for the blackfly and its natural enemies, the parasites can reduce the blackfly to such low levels that it is difficult to find at all. If and when the blackfly starts making a comeback, the parasites are there to increase their populations at the same time. Where the blackfly is eradicated by insecticides, there will be nothing to stop the pest the next time it is accidentally reintroduced into the area.

Another issue to consider with insecticide application is the secondary effects. Will the chemical use eliminate the pest but harm at the same time many natural enemies of other pests? That was just the concern that had Smith traveling repeatedly to the Rio Grande Valley in Texas over the years. Scientists now are getting a better understanding of the complex ecosystems that all forms of life inhabit and the many unintended and unpredicted effects that arise from harming one part of that ecosystem.

The business of chemical control of insects continues to be a multi-billion dollar industry. The war between biological control and insecticides, however, has found a sort of truce with the introduction of a new concept called IPM or Integrated Pest Control. Entomologists working in insect control are trying to integrate all forms of control in a scientific manner. Under this approach chemical control is used in a much narrower and tightly controlled and monitored way that will also allow biological control efforts to have maximum effect. New methods that modern science has come up with such as genetic modification and improved resistance are thrown into the mix.

Certainly Herbert Smith and his contemporaries in the biological control field were no environmental zealots. What they wanted was something that worked to improve people's lives. How well new ideas like IPM will work remains to be seen.

While Herb's work on the biological control of insect pests has not received the attention it has deserved, some experts in the field have taken note. One of Herb's colleagues, the UC Riverside entomologist Stanley Flanders, who assisted the blackfly work by providing guidance on shipping parasites via air express in wooden dugout containers, asked the Smith family for the loan of Herb's detailed field notes. Flanders used these notes to compile a review of Smith's blackfly work that was published in 1969.

University of California Riverside Professor Richard Stouthamer called the Mexican blackfly project "one of the most thorough biological control efforts ever." Texas A & M University Professor Julio Bernal called the project "a very important piece of biocontrol history and probably the best example (prior to cassava mealybug) of a large, complex, multi-national agency, and successful biocontrol effort in a poor country."

UC Riverside Professor Paul DeBach in his landmark book *Biological Control by Natural Enemies* reviewed Smith's work on the citrus blackfly and called it "one of the larger projects and great successes of modem times."

He pointed out that Smith's approach of importing all enemies of a pest was a key factor in achieving success across a large and diverse country such as Mexico.

In all Herb brought back eight parasite and two predator species to try against the blackfly. Of these, four became established and three became dominant in a complex interplay that developed over changing weather events over many years.

Someone hearing just the bare outlines of Herb's life, of the years spent in the field away from family, might wonder if perhaps he was one of those husbands and fathers who just preferred the carefree single life on the road. Anyone looking more closely at Herb's early days, when he was able to keep his family and work in the same orbit, and his warm and frequent letters home in his later years will realize that was far from the case.

He was a man who started work in a field that he believed would benefit mankind and only discovered later that agricultural politics in the US would prevent him from using his skills there. The restrictions on biological control work in the states meant that he had little choice if he was going to continue to support his family economically. As his daughter Janet explained, Herb did have a hard life.

The question of whether the efforts of Herbert Smith in controlling the citrus blackfly were worth the costs involved has no simple answer. The same question may be raised about the lives of many scientists and those in other lines of work who undertake a difficult task at great cost. His immediate family, who like Herb himself suffered the effects of long-term separation, are perhaps the only ones qualified to answer whether the personal sacrifices were worth what the world gained. There is little doubt however that all those who benefited from his efforts would answer with a clear and unhesitating yes.

Notes and Acknowledgments

The story of Herbert Smith would have been difficult if not impossible to tell without the detailed letters he wrote home to his family, and without his family preserving them. While Smith's detailed scientific reports describe the facts of his work, his letters present a different side of scientific work, the human dimension of doing a difficult undertaking under trying circumstances while being separated from home and family.

I would like to especially thank Herb's daughters Janet Rosier, Pat Stevens and Lucienne Robinson, who have collected and preserved Herb's letters from the field and shared photographs and documents and other information about Herb's life and work. Lucienne spent many hours organizing and transcribing the letters, which greatly enabled this research. I would also like to thank Catherine Deprez, Dr. Jeff Stevens, Chuck Robinson, Lee Smith, Pauline Lewis and David S. Lewis for sharing memories of the life of Herbert Smith and his career.

This effort was also greatly aided by Elliott Brichford and Lana Kitcher who did remarkable work examining tens of thousands of largely unindexed pages of U.S. Department of Agriculture files to find those that pertained to Herb and his work. These historic files, stored underground in old salt mines in Kansas, were airfreighted back to the National Archives in College Park, Maryland for study. These two talented and determined researchers recovered several thousand pages of important documents and letters that were critical in telling this story. Thanks also to the National Archives and Records Administration (NARA) staff who helped in making these documents available: Tim Nenninger, Richard Peuser, Theresa Mellon, Ann Cummings, Cindy Cutright and Eugene Morris.

Dr. Ru Nguyen of the Florida Department of Agriculture and Consumer Services, Division of Plant Industry provided helpful information about the citrus blackfly and biological control. He and his assistant Darleen George-Hill kindly provided photographs of the blackfly.

Others who provided helpful information or insights included Professor Julio S. Bernal (Department of Entomology, Texas A&M University, College Station), Dr. Hugo César Arredondo Bernal (Centro Nacional de Referencia de Control Biológico), Dr. Lynn M. LeBeck (director of the Association of Natural Bio-Control Producers), Robert Sawyer, Patti Fleschner, Jan Dietrick (Rincon-Vitova Insectaries), Professor Fred Legner (University of California, Riverside), and Professor Mark Hoddle (University of California, Riverside).

Herb Smith had a close and positive relationship with the Mexican agronomists and the Mexican people he was sent to help. Herb was a quiet and hard-working man, easy to get along with. But he could also be determined, stubborn really, when confronted with forces that interfered with what he was trying to accomplish. His relations with some of his colleagues in the U.S. Department of Agriculture, a large bureaucratic organization, and especially the other American entomologists working in Mexico were frequently difficult. At times they left him free to do his work. At other times they tried to rein him in. The fact that most of the department entomologists dealt mainly with insecticides – "spray men" as Smith called them – tended to make his work much more difficult.

As a result, his letters home frequently included strong criticism of some of the Americans he had to deal with. I have used a pseudonym for the entomologist identified in this story as Mason, as the man was working in a support role and had been sent down to Mexico without any specific training in the work. But I have left stand Herb's comments about those who were in a supervisory or equal position. Science is after all a bit of a wrestling match of competing ideas – and may the best idea win. In the case of the citrus blackfly in Mexico, it was the ideas and work of Herb Smith that succeeded.

Selected Bibliography

PRIMARY SOURCES

Flanders, Stanley E. (1969). Herbert D. Smith's observations on citrus blackfly parasites in India and Mexico and the correlated circumstances. Canadian Entomologist 101:467-475. Flanders reviewed Smith's notes and personal communications to provide a more in depth English language review of the scientific importance of the Mexican blackfly project. Unfortunately Smith's original notes were discarded by University of California - Riverside following the demise of the Biological Control Department.

Foreign Plant Introduction Quarterly Reports (1949-1952). Record Group 7: Records of the Bureau of Entomology and Plant Quarantine, National Archives, College Park, MD. Quarterly reports filed by Herbert D. Smith and colleagues.

Haeussler, G. J. and Harry L. Parker (1959). Herbert Dwight Smith, 1894-1959. Unknown publication, page 219.

Hill, C. C. and Herbert D. Smith (1931). *Heterospilus cephi* Rohwer, a parasite of the European wheat sawfly, *Cephus pygmeus* L. Journal of Agricultural Research 43:597-609.

Letter Files (1931-1959). Record Group 7: Records of the Bureau of Entomology and Plant Quarantine, National Archives, College Park, MD. Letters pertaining to the citrus blackfly, Herbert D. Smith, Theodore Gardner, Curtis Clausen, and A. C. Baker.

Robinson, Lucienne (2010-2014). Correspondence.

Rosier, Janet (2010-2014). Correspondence and interviews.

Smith, Herbert D. (1930). The bionomics of *Dibrachoides dynastes* (Foerster), a parasite of the alfalfa weevil. Annals of the Entomological Society of America 23:577-593.

Smith, Herbert D. (1932). *Phaeogenes nigridens* Wesmael, an important ichneumonid parasite of the European corn borer. Technical Bulletin 331. US Department of Agriculture, Washington, DC.

Smith, Herbert D. and Arlo M. Vance (1933). The larval head of parasitic Hymenoptera and nomenclature of its parts. Annals of the Entomological Society of America 26(1):86-94.

Smith, Herbert D. (1936). Le syrphide *Xanthadrus comtus* Harris predateur de la chenille processionaire du pin (*Cnethocampa pinivora* Tr.). Soc. Ent. de France Bul. 41:328-329.

Smith, Herbert D. (1943) Citrus Blackfly on the West Coast of Mexico and Attempts to Control it by importation and colonization of its most important natural enemy, *Eretmocerus serius* silver - June to August 1943. Trip Reports of Entomological Investigations. Record Group 7: Records of the Bureau of Entomology and Plant Quarantine, National Archives, College Park, MD. Herbert Smith's report on the Mexican blackfly situation in 1943.

Smith, Herbert. D. (1945). La "Mosca Prieta" de los citricos en la costa occidental de Mexico y la importacion y colonizacion de *Eretmocerus serius* Silv. para su control. Fitofilo 4:67-103. Smith's lengthy report in Spanish on the progress in control of the citrus blackfly in Mexico with parasites Smith had collected in Asia.

Smith, Herbert D. (1947). Scouting in Cuba for Catana clauseni Chapin: A Predator of the Citrus Blackfly. Bureau of Entomology and Plant Quarantine, Washington, D,C.

Smith, Herbert D. (1958). Las interelaciones de los enemigos naturales de la mosca prieta de los cítricos en México. Fitofilo 11:31-36. An updated Spanish language report on the blackfly program in Mexico.

Smith, Herbert D. Collected Letters, 1943 - 1959. Private Smith family collection. Letters written by Herbert Smith to his family, often with detailed accounts of his field work

Smith, Herbert D., Harry L. Maltby, Eleazar Jimenez Jimenez (1964). Biological control of the citrus blackfly in Mexico. USDA-ARS. Technical Bulletin No. 1311: 1-30. A posthumous report on the blackfly project in Mexico based on Smith's notes and reports, along with updates covering several years after Smith's death.

SECONDARY SOURCES

Abbas, H. M., M. S. Khan & H. Haque. 1955. Black fly of citrus (*Aleurocanthus woglumi* Ashby) in Sind and its control. Agric. Pakistan 6: 5-23.

Abbot, W. W. (1997). The Papers of George Washington. Confederation Series, January-September 1788. Charlottesville and London: University Press of Virginia, 358.

Afzal Husain, M., and Abdul Wahid Khan (1945). The Citrus Aleurodidae (Homoptera) in the Punjab and their control. Memoirs of the Entomological Society of India, no. 1:26-29.

Altieri, M. A. and C. I. Nichols. (1999). Classical biological control in Latin America, pp. 975-991. In Bellows, T. S. and T. W. Fisher (eds.). Handbook of Biological Control. Academic Press, San Diego, California, USA.

Ashby, S. F. (1915) Notes on diseases of cultivated crops observed in 1913-14. Jamaica Dept. Agr. Bull. 2:290-327.

Baker, W. A., W. C. Bradley and C A. Clark (1949). Biological Control of the European Corn Borer in the United States. U. S. Department of Agriculture Technical Bulletin 983. GPO, Washington D.C.

Battling citrus blackflies. (1990). Organic Gardening 37(2), 10.

Bedford, E. C. B., and E. D. Thomas (1965). Biological control of the citrus blackfly *Aleurocanthus woglumi* (Ashby) (Homoptera: Aleyrodidae) in South Africa. J. Entomol. Soc. S. Afr. 28(1):117-132.

Beingolea, D. O. (1996). El control biológico de plagas en el Perú: avances, perspectivas, aplicación y limitaciones, pp. 71-91. In Zapater, M. C. (ed.). El Control Biológico en América Latina. International Organization for Biological Control, Buenos Aires, Argentina.

Berger, M. Old Key West awakes. New York Times Magazine, March 21, 1937.

Berry, N. O. (1951). Progress report on the citrus blackfly surveys in Mexico. Proc. Rio Grande Valley Hort. Soc. 5:81-82.

Bess, H. A., Robert Van Den Bosch, and Frank Haramoto (1950). Progress and Status of Two Recently Introduced Parasites of the Oriental Fruit Fly, *Dacus dorsalis* Hendel, in Hawaii. Proceedings, Hawaiian Entomological Society 14(1):29-33.

Blackfly eradication program under way in Broward County. (1976). Fla. Grower and Rancher 68:19-20.

Brown, A. C. (1937). Report of the grove inspection department. Eleventh Biennial Report (1934-1936), 15-27. Gainesville: State Plant Board of Florida.

Bruner, S. C. (1922). Es muy danina la mosca prieta? Notas entomologicas. Rev. Agr. Com. y. Trab. [Cuba] 5(4): 6-8.

Bueno, V.H.P., J.C. van Lenteren (2002). The popularity of augmentative biological control in Latin America: history and state of affairs. Proceedings of the 1st International Symposium on Biological Control of Arthropods, Honolulu, Hawaii, 14-18 January 2002 pp. 181 (Ed: Driesche, R.G. van) United States Department of Agriculture, Forest Service, Washington DC.

Bueno, V. H. P. (1999). Protected cultivation and research on biological control of pests in greenhouses in Brazil. Bulletin of the International Organization for Biological Control, WPRS 22(1):21-24.

Burke, J. H. (1967). The commercial citrus regions of the world. In The citrus industry, Vol. 1, History, world distribution, botany, and varieties, ed. W. Reuther, H. J. Webber, and L. D. Batchelor, 40-189. Berkeley: Univ. Calif. Div. Agric. Sci.

Caffrey, D. J. and L. H. Worthley (1927). The European Corn Borer: Its Present Status and Methods of Control. Us Department of Agriculture Farmers Bulletin No. 1548.

Charles Valentine Riley Papers. Special Collection. National Agricultural Library, Beltsville, Maryland.

Chavez Torres, H. A. (1980). Biological control of the citrus blackfly (*Aleurocanthus woglumi* Ashby) by *Prospaltella opulenta* Silv. in central-western Venezuela with a review of the pest's invasion of the western hemisphere and suppression by introduced parasites. Ph.D. dissertation. University of Florida, Gainesville.

Cherry, R., and S. Pastor, Jr. (1980). Variations in population levels of citrus blackfly, *Aleurocanthus woglumi* (Horn.: Aleyrodidae) and parasites during an eradication program in Florida. Entomophaga 25:365-68.

Citrus blackfly moves into Palm Beach County. (1976). Plant Ind. News 17(4):25.

Citrus blackfly returns. (1976). Plant Ind. News 17(3):3-5.

Citrus blackfly: Florida eyes eradication. (1976). Calif. Citrogr. 62(2):43-44.

Citrus Fruits: Production, Farm Disposition, Value and Utilization of Sales, Crop Seasons 1909-10 to 1943-44. U. S. Department of Agriculture, Bureau of Agricultural Economics. Washington, D. C., October 1945.

Citrus trees threatened: Government warns of invasion by *Aleurocanthus woglumi*. New York Times Feb 7, 1921, Page 23.

Clausen, C. P. (1934). The natural enemies of Aleyrodidae in tropical Asia. Philippine J. Sci. 53: 253-65.

Clausen, C. P. (1936). Insect parasitism and biological control. Ann. Entomol. Soc. Amer. 24:201-23.

Clausen, C. P. (1937). Citrus Blackfly in the Bahamas. Trip Reports of Entomological Investigations. Record Group 7: Records of the Bureau of Entomology and Plant Quarantine, National Archives, College Park, MD.

Clausen, C. P. (1956). Biological control of insect pests in the continental United States. US Dept. Agr. Tech. Bull. 1139.

Clausen, C. P. (1978). Introduced parasites and predators of arthropod pests and weeds. U.S. Dept. Agric. Handbk. 480, 30-34.

Clausen Heads Biological Control. San Bernardino County Sun, July 11, 1951, page 16.

Clausen, C. P. and P. A. Berry (1933). The Citrus Black fly in Asia and the Importation of Its Natural Enemies into Tropical America. USDA Technical Bulletin No. 320. US GPO, Washington, DC.

Cooper, J. F., C. C. Plummer, and J. G. Shaw (1950). The citrus blackfly situation in Mexico. J. Econ. Entom. 43:767-773.

Corn Borer Menace. New York Times, July 24, 1921.

Coulson, J. R., P. V. Vail, M. E. Dix, D. A. Nordlund, and W. C. Kauffman [eds]. (2000). 110 years of biological control research and development in the United States Department of Agriculture. USDA ARS Admin. Rep. 2000-1.

Current Notes (1922). Mr. Harry L. Parker, attached to the corn borer investigations, sailed for France on January 17. Journal of Economical Entomology 15:188.

Deaths Near 1500 in Mexico storm. Pacific Stars and Stripes. November 3, 1959. p. 31.

DeBach, Paul (1964). Successes, trends, and future possibilities. In Biological control of insect pests and weeds, ed. P. DeBach, 673-713. N.Y.: Reinhold.

DeBach, Paul (1974). Biological control by Natural Enemies. Cambridge University Press, London.

DeBach, P., and B. Bartlett (1951). Effects of insecticides on biological control of insect pests of citrus. J. Econ. Entom. 44:372-383.

DeBach, Paul (1976). Biological Control Success in Mexico. Citrograph 62(2):58-59. Reprint of article about Herbert Smith in DeBach's book Biological Control by Natural Enemies,

Delgado, D and A. Garay (1943). Control biológico de la mosca prieta de los cítricos en la republica. Fitofilo 5: 1-17.

Dietrick, Jan (2011), Correspondence about Stanley Flanders and the history of biological control in California.

Dietrick, Jan and Everett J. "Deke" Dietrick. The First Commercial Natural Enemy Venture. Accessed online on 1-2-2015 at http://www.rinconvitova.com/history%20BC.htm.

Dietz. H. F., and Zetek, J. (1920). The black fly of citrus and other subtropical plants. U. S. Dept. Agr. Bull. 885, 55 p.

Dominguez, E. R. (1996). Control biológico de plagas agrícolas en México, pp. 55-62. In Zapater, M. C. (ed.). El Control Biológico en América Latina, International Organization for Biological Control, Buenos Aires, Argentina.

Dowell RV, Cherry RH, Fitzpatrick GE, Reinert JA, Knapp JL. (1981). Biology, plant-insect relations, and control of the citrus blackfly. Florida Agricultural Experiment Station Bulletin 818: 1-48.

Dowell, Robert V.; Fitzpatrick, George E.; Reinert, James A. (1979). Biological Control of Citrus Blackfly in Southern Florida. Environmental Entomology 8(4):595-597.

Edwards, W. H. (1932). Importation into Jamaica of a parasite (*Eretmocerus serius* Silv.) of the citrus black fly (*Aleurocanthus woglumi* Ash.). Jamaica Dept. Sci. & Agric. Ent. Bull. 6. 12 p.

Eleazar, J. M. (1950). Control work on citrus black fly is studied in old Mexico. Florence Morning News, December 17, 1950, page 4-A.

Enkerlin, D. (1976). Some aspects of the citrus blackfly (*Aleurocanthus woglumi* Ashby) in Mexico. Proceedings: Tall Timbers Conference on Ecological Animal Control by Habitat Management (Tallahassee, Fla.) 6:65-76

Fight citrus black fly. New York Times, April 10, 1921.

Fight starts in citrus black fly south of border. Fresno Bee, November 16, 1947.

French, J. V., R. L. Meagher, K. L. Esau (1990). Release of two parasitoid species for biological control of citrus blackfly in south Texas. J. Rio Grande Valley Hort. Soc. 43:23-27.

Fruit fly infestation is feared. Santa Cruz Sentinel, Feb 21, 1954, page 4.

Funds Urged for Citrus Fly Fight. Lubbock Morning Avalanche, April 22, 1949, page 27.

Gahan, Arthur Burton (1933). The Serphoid and Chalcidoid Parasites of the Hessian Fly. Miscellaneous Publication 174. US Department of Agriculture, Washington DC.

Garcia, F. (1996). El control biológico aplicado en Colombia, pp. 31-33. In Zapater, M. C. (ed.). El Control Biológico en América Latina. International Organization for Biological Control, Buenos Aires, Argentina.

Geigera, Flavia Geiger, Jan Bengtsson, Frank Berendse et al. (2010). Persistent negative effects of pesticides on biodiversity and biological control potential on European farmland. Basic and Applied Ecology 11(2):97–105.

Gerald, Sam (1966). Two interceptions alert Valley to peril from citrus black fly. Brownsville Herald, August 21, 1966, page 16A.

Gomes, J. (1962). Histórico do combate biológico no Brasil. Boletim do Instituto de Ecologia e Experimentação Agricolas 21: 89-97.

Gonzalez, R. H. and S. Rojas.1966. Estudio analitico del control biologico de plagas agricolas en Chile. Agricultura Tecnica 26: 133-147.

Gowdey. C. C. (1921). The citrus black fly, (*Aleurocanthus woglumi* Ashby). Jamaica Dept. Agr. Ent. Circ. 3 of 1021, 11 p.

Gunter, F. (1954). Black fly control in Mexican groves. Citrus Leaves 54:20.

H. D. Smith, in Mexico: Rutherford Man, with U.S. Agency, Dies Suddenly. Newark Evening News, January 23, 1959.

Hagen, K. S. and J. M. Franz (1973). A history of biological control, pp. 433-476. In Smith, R. F., T. E. Mittler, and C. N. Smith (eds.). History of Entomology. Annual Reviews Inc., Palo Alto, California, USA.

Harry Scott Smith, Biological Control, Riverside (1959). University of California, In Memoriam.

Hart WG, Selhime A, Harlan DP, Ingle SJ, Sanchez-R M, Rhode RH, Garcia CA, Caballero J, Garcia RL. (1978). The introduction and establishment of parasites of citrus blackfly, *Aleurocanthus woglumi* in Florida (Hem.: Aleyrodidae). Entomophaga 23: 361-366.

Henshaw, Samuel (1890). Bibliography of the More Important Contributions to American Economic Entomology: The more important writings of Benjamin Dana Walsh and Charles Valentine Riley. U.S. Government Printing Office, Washington, D.C.

Herbert D. Smith, 65, Dies: Agricultural Department Expert. The Herald News, January 23, 1959.

Hernández, J. (1996). El control biológico en Costa Rica, pp. 35-40. In Zapater, M. C. (ed.). El Control Biológico en América Latina. International Organization for Biological Control, Buenos Aires, Argentina.

Heu, Ronald A. and Walter T. Nagamine (2001). Citrus blackfly. New Pest Advisory 99-03. State of Hawaii Department of Agriculture, Honolulu, Hawaii.

Hoelmer, Kim A. and J. Kenneth Grace (1989). Citrus Blackfly. In Eradication of Exotic Pests: Analysis with Case Histories. Yale University Press: New Haven, p. 147-165. .

Holler, T. C. and J. R. Brazzel (1978). Colonization of citrus blackfly parasites in the lower Rio Grande Valley of Texas. Citrus Industry 59:42-46.

House Group Vote Funds To Fight War On Citrus Blackfly. Fresno Bee, March 16, 1948, page 3.

Hunter, Brooke (2006). Creative Destruction: The Forgotten Legacy of the Hessian Fly. In The Economy of Early America: Historical Perspectives & New Directions, edited by Cathy D. Matson. Penn State Press.

In memorium (2000). Noel Louis Hilmer Krauss (1912-1996). Proceedings of the Hawaiian Entomological Society 34 : v-viii.

Jimenez, E.J. (1958). El *Syntomosphyrum indicum*, un enemigo mortal de la mosca de la fruta (*Anastrepha ludens* Loew.). Fitofilo 11(21):25-30.

Jimenez, E.J. (1960). Estado actual de la mosca prieta de los citricos y adelantos en las trabajos de campo y laboratory. Fitofilo 13(28):41-48.

Jimenez, E.J. (1961). Situation actual de la mosca prieta (*Aleurocanthus woglumi* Ash.), de los citricos en Mexico. Fitofilo 14(32):39-45.

Jimenez, E.J. (1963). Control biológico de la mosca prieta en México. Fitofilo 16: 6-41.

Jimenez, E.J. (1964). Concentration de datos de los trabajos desarrollados contra la mosca prieta de los citricos durante los anos de 1959 a 1964. Fitofilo 17 (43):46-48.

Jimenez, E.J. (1965). Situación de la mosca prieta de los cítricos en México. Fitofilo 18: 37-40.

Jimenez, E.J. & T. R. Carillo (1968). La campaña contra la mosca prieta de los cítricos, *Aleurocanthus woglumi* Ashby. Fitofilo 21(60):23-28.

Jimenez, E.J. (1970). Importancia economica del combate biologico de la mosca prieta de los citricos en Mexico. Fitofilo 23(65):9-11.

Jimenez, E.J. (1971). Comportamiento de los enemigos naturales de la mosca prieta de los citricos en la republica Mexicans. Fitofilo, 24(66):2-6.

Jimenez, E.J. (1972). Trampeo de la mosca prieta de los citricos (*Aleurocanthus woglumi* Ash.). Folia Ent. Mex. 23/24:59.

Kamal, M. (1951). Biological control projects in Egypt, with a list of introduced parasites and predators. Bull. Soc. Fouad Ier. Ent. 35:205-20.

Ketner, C. F., and J. G. Rosier (1978). Citrus blackfly controlled biologically. Texas Agr. Prog. 2:19-20.

Kogan, M. (1998). Integrated pest management: historical perspectives and contemporary developments. Annu Rev Entomol. 43:243-70.

Kuwana, I., and Ishii, T. (1927). On *Prospaltella smithi* Silv. and *Cryptognatha* sp., the enemies of *Aleurocanthus spiniferus*, imported from Canton, China. Jour. Okitsu Hort. Soc. No. 22. p. 77-80.

Likes, Ken (1947). Insects to spearhead Navy attack. Danville Virginia Bee, November 3, 1947.

MacFarlane, David L. (1948). The UNRRA Experience in Relation to Developments in Food and Agriculture. Journal of Farm Economics 30(1):69-77.

Martinez, Nancy Boscán de, Jorge Terán B., and Francis Geraud (1979). Enemigos naturales de la mosca prieta de los citricos (*Aleurocanthus woglumi* Ashby) en Venezuela. Agronomía Tropical. 29(5): 453-458.

Mateos, D. L. A. (1963). Control biologico de la mosca prieta en Mexico. Fitofilo 16(37):5-41.

Meagher, R. L., & French, J. V. (2004). Augmentation of parasitoids for biological control of citrus blackfly in southern Texas. The Florida Entomologist, 87(2), 186-193.

Mexico Fights Threat of Epidemic after Hurricane that Killed 2,000. Ogden Standard-Examiner, November 2, 1959. p. 8.

Monthly Letter of the Bureau of Entomology, United States. Various Issues. United States Bureau of Entomology, Washington DC.

Morrill. A. W., and E. A. Back (1912). Natural control of white flies in Florida. U. S. Dept. Agr. Bur. Ent. Bull. 102:78pp.

Newell W, Brown AC. (1939). Eradication of the citrus blackfly in Key West, Fla. Journal of Economic Entomology 32: 680-682.

Newell, W. (1918). The "Blackfly," a serious menace to citrus. Calif. Citrograph. 3(9):209, 216-7.

Nguyen, Ru, Brasil JR, Poucher C. (1983). Population density of the citrus blackfly, *Aleurocanthus woglumi* Ashby (Homoptera: Aleyrodidae), and its parasites in urban Florida in 1979-81. Environmental Entomology 12: 878-884.

Nguyen, Ru. (1987). *Encarsia opulenta* (Silvestri) a parasite of *Aleurocanthus woglumi* Ashby (Hemiptera: Atherinidae). Florida Department of Agriculture & Consumer Services, Division of Plant Industry. Entomology Circular 301. 2 p.

Nguyen, Ru (2011). Correspondence about and photos of the citrus blackfly in Florida.

Quezada, J. R. (1974). Biological control of *Aleurocanthus woglumi* (Homoptera: Aleyrodidae) in El Salvador. Entomophaga 19:243-254.

Quien es quien en control biologico en Mexico: Directorio de especialistas, instituciones y laboratorios de organismos beneficos. Website of Sociedad Mexicana de Control Biológico http://www.smcb-mx.org.

Rao, V. P. (1969). India as a source of natural enemies of pests of citrus. Proceedings of the First International Citrus Symposium 2:785-792.

Riley, Charles Valentine (1870). Second annual report on the noxious, beneficial, and other insects of the State of Missouri, made to the State Board of Agriculture, pursuant to an appropriation for this purpose from the legislature of the State. Missouri State Board of Agriculture, Jefferson City, Missouri.

Riley, Charles Valentine (1885). On the Parasites of the Hessian Fly. Proceedings of the United States National Museum. P. 413-422.

Riley, Charles Valentine (1887). Report of the entomologist. U.S. Government Printing Office. Washington, D.C.

Sandford, H. L. (1921). Citrus Blackfly in Florida. Trip Reports of Entomological Investigations. Record Group 7: Records of the Bureau of Entomology and Plant Quarantine, National Archives, College Park, MD.

Sauls, Julian W. (2008). The Texas Citrus Industry. http://aggie-horticulture.tamu.edu/citrus/l2286.htm.

Scientists pester blackfly with insects. Brazosport Facts, January 21, 1989.

Semi-monthly Narrative: Mexican Fruit Fly and Citrus Blackfly Control Project (1952). Technical Program Records. Record Group 7: Records of the Bureau of Entomology and Plant Quarantine, National Archives, College Park, MD.

Sierra, Jose Luis. Mexico Dispatch: A Town in the Heart of Narco Country, Pretending It Isn't. PBS Newshour,

November 30, 2010. http://www.pbs.org/newshour/rundown/mexico-dispatch-3/.

Silvestri, F. (1926). Descrizione di tre specie di Prospaltella e di una Encarsia parassite di Aleurocanthus. EOS, Eev. Espan. Ent. 2. (2-3) : 179-189.

Silvestri, F. (1927). Contribuzione alla conoscenza degli Aleurodidae (Insecta-Hemiptera) viventi su Citrus in Extremo Oriente e dei loro parassiti. Boll. Lab. Zool. Gen. Agr. Portici 21:1-60.

Silvestri, F. (1927). Sobre la mosca prieta. Dept. Agr. [Costa Rica]. Bol. Fomento 6:16-18.

Simons, Savilla Millis (1944). "U.N.R.R.A. on the Threshold of Action." The Social Service Review 18(4):433-443.

Slawter, Clayton R, (1932). Bug eats bug to save American farms. Popular Science, July 1932, p. 44-45, 100-110.

Stouthamer, R. (2003). The Use of Unisexual Wasps in Biological Control. In Quality Control and Production of Biological Control Agents: Theory and Testing Procedures, edited by J. C. van Lenteren. CABI Publishing, Wallingford, UK, pages 93-115.

Summy KR, Gilstrap FE, Hart WG, Caballero JM, Saenz I. (1983). Biological control of citrus blackfly (Homoptera: Aleyrodide) in Texas. Environmental Entomology 12: 782-796.

The blackfly question: Eradicate or biocontrol? 1979. Fla. Grower and Rancher 72(4):12-13.

The citrus blackfly fight in the lower Rio Grande Valley. (1975). Citrus Ind. 56(7):8,10-11.

Thompson, W. R., and Parker, H. L. (1928). The European Corn Borer and its Controlling Factors in Europe. U. S. Dept. Agr. Tech. Bul. 59, 63 p.

Trujillo-Arriaga, Francisco Javier (2013). Semblanza del Ing. Eleazar Jimémez Jiménez (Biographical Sketch of Eleazar Jimémez Jiménez). Vedalia 14(1):3-4.

Two minutes to abandon ship! SS Washington in Stand-off with U-Boat! http://usmm.org/washington.html.

U.N.R.R.A. 1945-1947: The End of a Chapter. The World Today Vol. 3, No. 8 (Aug., 1947), pp. 370-374.

U.N.R.R.A.'s Final Report. The Social Service Review, Vol. 23, No. 2 (Jun., 1949), p. 245.

Vail, P. V., J. R. Coulson, W. C. Kauffman, and M. E. Dix (2001). History of Biological Control Programs in the United States Department of Agriculture. American Entomologist 47(1):24-49.

Van Driesche, Roy G. and Thomas S. Bellows Jr. (1996). Biological Control. Springer.

Vance, Arlo (1932). *Microgaster tibialis* Nees as a Hymenopterous Parasite of *Pyrausta nubilalis* Hubn. in Europe. Annals of the Entomological Society of America 25(1):121-135.

Warner, K. D., C. Getz, S. Maurano, K. Powers (2009). An analysis of historical trends in classical biological control of arthropods suggests need for a new centralized database in the USA. Biocontrol Science and Technology 19(7): 675-688.

Wille, J. E. (1956). El control biológico de los insectos agrícolas en el Perú. Proceedings International Congress of Entomology 10: 519-523.

Williams, Trevor, Hugo C. Arredondo-Bernal, and Luis A. Rodríguez-del-Bosque (2013). Biological Pest Control in Mexico. Annual Review of Entomology 58: 119-140.

Wilson, Lyle C. C. L. Marlatt is in Charge Of Recruiting Army of Parasites Combat Agricultural Pests, Vidette-Messenger, Oct 1, 1930, page 8.

Woglum, Russell S. (1913). Report of a trip to India and the Orient in search of natural enemies of the citrus white fly. US Dept. Agr. Tech. Bull. 120.

Woglum, Russell S. and Harry S. Smith (1947). The citrus blackfly. Calif. Citrograph 32:412-415.

Woglum, Russell S. (1948). The California-Arizona citrus growers campaign against the citrus black fly on the west coast of Mexico. The Exchange Pest Control Circular, August 1948. California Fruit Growers Exchange, Los Angeles.

Woglum, Russell S., Harry S. Smith and Curtis Clausen (1952). The citrus blackfly in northwest Mexico. Calif. Citrograph 37:356.

Woodbridge, George et al. UNRRA: The History of the United Nations Relief and Rehabilitation Administration, Vols I-III. New York: Columbia University Press, 1950.

Yepes, Victor (1957). Toll put at 55 in worst quake, Miami Daily News, July 29, 1957, p. 1.

Made in the USA
Monee, IL
23 September 2025

30044290R00163